SpringerBriefs in Mathematics

Series Editors

Krishnaswami Alladi
Nicola Bellomo
Michele Benzi
Tatsien Li
Matthias Neufang
Otmar Scherzer
Dierk Schleicher
Vladas Sidoravicius
Benjamin Steinberg
Yuri Tschinkel
Loring W. Tu
G. George Yin
Ping Zhang

SpringerBriefs in Mathematics showcases expositions in all areas of mathematics and applied mathematics. Manuscripts presenting new results or a single new result in a classical field, new field, or an emerging topic, applications, or bridges between new results and already published works, are encouraged. The series is intended for mathematicians and applied mathematicians.

For further volumes:
http://www.springer.com/series/10030

Juan Pablo Pinasco

Lyapunov-type Inequalities

With Applications to Eigenvalue Problems

 Springer

Juan Pablo Pinasco
Departamento de Matematica
Universidad de Buenos Aires
Buenos Aires, Argentina

ISSN 2191-8198   ISSN 2191-8201 (electronic)
ISBN 978-1-4614-8522-3   ISBN 978-1-4614-8523-0 (eBook)
DOI 10.1007/978-1-4614-8523-0
Springer New York Heidelberg Dordrecht London

Library of Congress Control Number: 2013947680

Mathematics Subject Classification (2010): 34L15, 34B05, 34B15, 34C10, 35P30

© Juan Pablo Pinasco 2013

This work is subject to copyright. All rights are reserved by the Publisher, whether the whole or part of the material is concerned, specifically the rights of translation, reprinting, reuse of illustrations, recitation, broadcasting, reproduction on microfilms or in any other physical way, and transmission or information storage and retrieval, electronic adaptation, computer software, or by similar or dissimilar methodology now known or hereafter developed. Exempted from this legal reservation are brief excerpts in connection with reviews or scholarly analysis or material supplied specifically for the purpose of being entered and executed on a computer system, for exclusive use by the purchaser of the work. Duplication of this publication or parts thereof is permitted only under the provisions of the Copyright Law of the Publisher's location, in its current version, and permission for use must always be obtained from Springer. Permissions for use may be obtained through RightsLink at the Copyright Clearance Center. Violations are liable to prosecution under the respective Copyright Law.

The use of general descriptive names, registered names, trademarks, service marks, etc. in this publication does not imply, even in the absence of a specific statement, that such names are exempt from the relevant protective laws and regulations and therefore free for general use.

While the advice and information in this book are believed to be true and accurate at the date of publication, neither the authors nor the editors nor the publisher can accept any legal responsibility for any errors or omissions that may be made. The publisher makes no warranty, express or implied, with respect to the material contained herein.

Printed on acid-free paper

Springer is part of Springer Science+Business Media (www.springer.com)

To Ceci, Fede, and Selva

Preface

I used to think that the Sturm–Liouville theory of second-order ordinary differential equations was one of the most beautiful areas of mathematics. Its simplicity, together with the power of the comparison and oscillation theorems, shed a different light on second-order ordinary differential equations. However, while reading a transcription of a talk of G.C. Rota, I realized something: there are many interesting problems, both of theoretical and applied origin, that cannot be analyzed with the Sturmian tools.

Take the unit ball in R^N: just the simple reduction to polar coordinates introduces the coefficient r^{N-1}, which vanishes at the origin and is bounded above by 1, for all N. Moreover, Bessel, Hermite, Legendre, ..., almost all the special families of functions that appear as eigenfunctions of second-order ordinary differential operators, are indeed eigenfunctions of singular or degenerate operators, and the Sturmian arguments fail. What can we do now?

If we write the Sturmian bounds in modern notation, we are using the L^∞ norm of the weight, and what happens if we change it to another norm, say L^1? Indeed, the answer is known, and it is related to the stability of solutions of second-order differential equations, a problem studied by Lyapunov almost 150 years ago. He introduced an integral condition that the weight must satisfy in order to guarantee stability. However, he never proved Lyapunov's inequality. Later, Borg, Hartman, Krein, and other mathematicians working on stability gave his name to this kind of Sturmian bound with an L^1 norm.

However, unbounded domains still present a difficulty, since Lyapunov's inequality includes the length of the interval on which the problem was studied. We might decide to ignore this problem, dismissing it as a hifalutin theoretical question. But not so fast! It was, in fact, a legitimate question, inspired by quantum mechanics and related to the number of bound states of the Schrödinger equation. Ordinary differential equations on unbounded intervals were studied in the 1950s and 1960s by Jost, Pais, Bargmann, Calogero, Cohn, and Nehari (the only one who was not thinking of quantum-mechanical problems), among several others. They obtained beautiful inequalities, involving different norms of the coefficients.

And in the last twenty years, many mathematicians have extended those results to a variety of settings, including p-Laplacian operators, ordinary differential equations in Orlicz spaces, N-dimensional problems, and systems.

I designed this book as a guided tour through those results, together with their applications to eigenvalue problems, presenting full proofs and extensions of those inequalities, and showing the less-traveled paths, suggesting directions for future work. I tried to include in the references all the relevant papers on this subject, and I apologize here for the inevitable omissions.

I wish to thank several people who contributed directly or indirectly to this book: P. Amster, J.M. Castro, P. De Nápoli, J. Fernández Bonder, and A. Salort. Also, I wish to thank the people at UCo-CEMIC, Buenos Aires, for their hospitality, and the financial support from Universidad de Buenos Aires and CONICET.

Buenos Aires, Argentina   Juan Pablo Pinasco

Contents

Symbols and Notation

Throughout this work,

- The letters a, b, c, d denote real numbers.
- p, q denote real numbers greater than one.
- p' denotes the Hölder conjugate of p, $p' = p/(p-1)$.
- The letters i, j, k, m, n denote a positive integer.
- λ denotes a real parameter, usually an eigenvalue of some differential operator.
- By c, C we denote positive constants, and we write $C(x, y, \ldots)$ whenever we need to stress the dependence of C on $x, y, \ldots$.
- R^N is N-dimensional Euclidean space, with $N \geq 1$.
- $\Omega \subset R^N$ is a bounded open set.
- $\partial\Omega$ is the boundary of Ω.
- u, v, w denote real-valued functions.
- Given $u : (a, b) \to R$, $u'(x)$ and $\frac{du}{dx}$ denote the derivative of u. Derivatives of higher orders are denoted by u'', $u^{(m)}$, $\frac{d^m u}{dx^m}$.
- Given $u : \Omega \to R$ with $\Omega \subset R^N$, ∇u denotes the gradient of u.
- $C_0^\infty(\Omega)$ is the space of C^∞ functions with compact support in Ω.
- If $1 \leq p \leq \infty$, $L^p(\Omega)$ denotes the usual Lebesgue space of measurable functions $u : \Omega \to R$, with the norm

$$\|u\|_p = \left(\int_\Omega |u|^p \mathrm{d}x \right)^{\frac{1}{p}}.$$

$$\|u\|_\infty = \mathrm{esssup}\{|u(x)| : x \in \Omega\}.$$

- $L^1_{\mathrm{loc}}(\Omega)$ denotes the set of measurable functions integrable on compact subsets of Ω.
- $W^{m,p}(\Omega)$, $W_0^{m,p}(\Omega)$, $1 \leq p < \infty$, $1 \leq m$ denote the usual Sobolev spaces.

Chapter 1
Introduction

Abstract In this chapter, we present four theorems that will be proved and generalized in the body of the book. Each one gives a lower bound for the first eigenvalue of a weighted second-order ordinary differential equation, and they involve different integrals of the weight. The main aim of the book is to prove these theorems and show the interplay among them. We also present several extensions to nonlinear equations, systems, and more general operators, which are briefly described here.

> We may regard the present state of the universe as the effect
> of its past and the cause of its future.
> An intellect which at a certain moment would know
> all forces that set nature in motion,
> and all positions of all items of which nature is composed,
> if this intellect were also vast enough to submit these data to analysis,
> it would embrace in a single formula
> the movements of the greatest bodies of the universe
> and those of the tiniest atom.
>
> —Pierre Simon Laplace, *A Philosophical Essay on Probabilities*

1.1 A Few Words About Four Theorems

In the mid 1800s, Laplace's dream was confronted with an unexpected difficulty: equations derived from Newton's second law,

$$-\frac{\mathrm{d}}{\mathrm{d}t}\left[p(t)\frac{\mathrm{d}u}{\mathrm{d}t}\right]+q(t)u(t) = f(t), \qquad t \geq 0,$$

together with initial conditions of position and velocity, $u(0) = u_0$, $u'(0) = u'_0$, could not be solved explicitly, despite the efforts of many mathematicians.

J.P. Pinasco, *Lyapunov-type Inequalities: With Applications to Eigenvalue Problems*,
SpringerBriefs in Mathematics, DOI 10.1007/978-1-4614-8523-0_1,
© Juan Pablo Pinasco 2013

This type of equations pervades several branches of physics, appearing in planetary dynamics, vibrations of inhomogeneous strings, heat conduction in one-dimensional objects, and propagation of one-dimensional waves, among many other problems.

Of course, the solutions for some particular cases were well known: Euler introduced the characteristic equation for constant-coefficient problems around 1730; Bessel generalized the radial solutions (introduced by Daniel Bernoulli) that we know now as Bessel functions around 1820; and the work of Fourier, combining series expansions and separation of variables, made it possible to solve many partial differential equations.

However, in order to apply the Fourier method for linear operators with nonconstant coefficients in $\Omega = [0,L] \times (0,\infty)$, we need to know the spectrum and eigenfunctions of second-order problems of the form

$$-\frac{d}{dx}\left[p(x)\frac{du}{dx}\right] + q(x)u(x) = \lambda w(x)u(x),$$

with some boundary condition induced by the boundary conditions of the original problem, such as the zero Dirichlet boundary conditions

$$u(0) = u(L) = 0$$

or the zero Neumann boundary conditions

$$u'(0) = u'(L) = 0.$$

Sturm, who was working as an assistant to Fourier and Liouville, studied this problem, and in the mid 1830s, these mathematicians obtained the core results of Sturm–Liouville theory:

Theorem 1.1. *Let us assume that p, w > 0, p, p', q, and w are continuous functions on a closed bounded real interval [0,L], and let us consider the eigenvalue problem*

$$-\frac{d}{dx}\left[p(x)\frac{du}{dx}\right] + q(x)u(x) = \lambda w(x)u(x),$$

with zero Dirichlet boundary conditions. Then:

1. There exists a sequence of real eigenvalues $\{\lambda_k\}_{k \geq 1}$ such that

$$\lambda_1 < \lambda_2 < \cdots < \lambda_n < \cdots \to +\infty.$$

2. To each eigenvalue λ_n there corresponds a unique (up to a normalization constant) eigenfunction u_n, which has exactly $n + 1$ zeros in $[0,L]$.
3. The normalized eigenfunctions form an orthonormal basis of some Hilbert space included in $L^2([0,L],w)$.

Clearly, point 3 of Theorem 1.1 was neither stated nor proved in this form, although Liouville proved the existence and convergence of generalized Fourier expansions in terms of the eigenfunctions for any reasonable function. We refer the interested reader to the beautiful article of Lützen and Mingarelli [79] for a careful historical and mathematical description of the work of Sturm and Liouville on this subject, and to the original articles of Sturm.

However, this is a pure existence result, without a clue as to the actual computation of the solutions. Sturm approached this problem in a different way in his memoir [107]:

> S'il importe de pouvoir déterminer la valeur de la fonction inconnue pour une valeur isolé quelconque de la variable dont elle dépend, il n'est pas moins nécessaire de discuter la marche de cette fonction, ou en d'autres termes, d'examiner la forme et les sinuosités de la courbe dont cette fonction serait l'ordonné variable, en prenant pour abscisse la variable indépendante.

Then he proved some oscillation and comparison results. We can summarize their results in the following theorems:

Theorem 1.2. *Let us assume that $p, w > 0$, p, p', and w are continuous functions on a closed bounded real interval $[0, L]$, and let u, v be two solutions of*

$$-\frac{d}{dx}\left[p(x)\frac{du}{dx}\right] = w(x)u(x).$$

Then the zeros of u and v are isolated, and between two zeros of u, there exists a unique zero of v.

Theorem 1.3. *Let us assume that $w_1 > w_2 > 0$, and $0 < p_1 < p_2$ and that p'_1, p'_2 exist and are continuous functions on a bounded real interval $[0, L]$. Let u and v be solutions of the equations*

$$-\frac{d}{dx}\left[p_1(x)\frac{du}{dx}\right] = w_1(x)u(x),$$

$$-\frac{d}{dx}\left[p_2(x)\frac{dv}{dx}\right] = w_2(x)v(x).$$

Then between two zeros of u, there exists at least one zero of v.

These theorems have interesting consequences for the relationships among eigenvalues of different problems. In particular, we get a simple bound for the first eigenvalue of the following problem in $[0, L]$:

$$\begin{cases} -\dfrac{d^2 u}{dx^2} = \lambda_1 w(x)u, \\[2mm] u(0) = 0 = u(L), \end{cases} \tag{1.1}$$

namely:

Theorem A. *Let us assume that* $0 \leq w \leq M$ *in* $(0,L)$, *and let* λ_1 *be the first eigenvalue of problem (1.1). Then*

$$\frac{\pi^2}{ML^2} \leq \lambda_1.$$

We can say that Theorem A gives a lower bound for λ_1 involving $\|w\|_\infty$, the $L^\infty(0,L)$ norm of w.

One could ask whether there exists a lower bound for λ_1 involving different norms of w. Indeed, for the L^1 norm, such a lower bound is a consequence of a result proved by Borg almost a century ago when he was trying to improve Lyapunov's stability theorem; see [9, 80]. We have the following theorem:

Theorem B (Lyapunov inequality). *Let us assume that w is a continuous function allowed to change sign in $(0,L)$, and let u be a nontrivial solution of*

$$-\frac{\mathrm{d}^2 u}{\mathrm{d}x^2} = w(x)u$$

satisfying $u(0) = 0 = u(L)$. *Then*

$$\frac{4}{L} \leq \int_0^L |w(x)|\mathrm{d}x.$$

In particular, the first eigenvalue of problem (1.1) can be bounded as follows:

$$\frac{4}{L\displaystyle\int_0^L |w(x)|\mathrm{d}x} \leq \lambda_1.$$

However, there are some problems that suggest the need for improvement in the previous inequalities.

The first is the asymptotic behavior of the eigenvalues. It is impossible to give a short list of those who have worked on this subject, but let us mention only Weyl [112], Courant [31], and Mark Kac [70]. For problem (1.1), it is well known that the nth eigenvalue λ_n behaves like

$$\lambda_n = \frac{\pi^2 n^2}{(\int_0^L w^{\frac{1}{2}}\mathrm{d}x)^2} + O(1)$$

as $n \to \infty$, so it is natural to ask whether there exists a lower bound involving an expression related to it. The answer for some weights is positive, and it is due to Nehari [88], who proved in 1959 the following inequality:

Theorem C. *Let $w \geq 0$ be a monotonic function, and let λ_1 be the first eigenvalue of problem (1.1). Then*

$$\frac{\pi}{2} \leq \lambda_1^{\frac{1}{2}} \int_0^L w^{\frac{1}{2}}(x)\mathrm{d}x. \tag{1.2}$$

This theorem can be extended to monotonic weights $w \in L^1$, and can be applied to more general weights. For example, let us mention briefly another problem that can be studied with the help of Theorem C: the linearized Kolodner rotating string equation,

$$-\frac{\mathrm{d}^2 u}{\mathrm{d}x^2} = \lambda w(x)\frac{u}{x}, \qquad x \in (0,L),$$

with mixed boundary conditions

$$u(0) = 0 = u'(L).$$

For bounded and strictly positive functions w, the weight $w(x)/x$ does not belong to L^1 or L^∞. When $w \equiv 1$, the problem can be solved using Bessel functions; see [71]. There are few results related to this problem; see the survey of Antman [3] and the work of Riddell [103] for more information about it. Let us mention that for this particular problem, an extension of Lyapunov's inequality due to Hartman and Wintner [63] enables us to deal with L^1_{loc} weights.

Another interesting problem is the location of zeros of solutions of second-order problems defined on $[0, \infty)$. The inequalities derived from Theorems A and B involve the length of the interval, and so they cannot be used in this case. This was the motivation behind the following theorem, proved in 1965 by Calogero [17] and in the same year by Cohn [28]:

Theorem C'. *Let w be a continuous and monotonic function, and let u be a solution of*

$$-u'' = w(x)u, \qquad x \in (0,\infty).$$

Suppose that $u(a) = u(b) = 0$, where $0 < a < b < \infty$. Then

$$\frac{\pi}{2} \leq \int_a^b w^{\frac{1}{2}}(x)\mathrm{d}x. \tag{1.3}$$

Clearly, Theorems C and C' are equivalent, the only difference being the way they are stated and the motivation behind them. As a consequence of Theorem C', Calogero and Cohn proved what is now called the Calogero–Cohn bound for the number of negative eigenvalues of one-dimensional Schrödinger operators, and extended the result to radial potentials in higher dimensions. Their result is equivalent to a related work of Hille [64] of the same year, which we will discuss later, although he never stated explicitly the previous inequality.

In the last 15 years, several works have appeared devoted to extensions and generalizations of those results, related to Schrödinger equations in waveguide domains and higher-order operators, mainly by Birman, Naimark, and Solomyak; see the works [6, 84, 105] and the references therein. Their proofs are very different from the previous ones, and the authors do not seem to have been aware of the results of Hille and Nehari.

On the other hand, Theorem C has a strong limitation: the monotonicity of the weight function. However, it is possible to derive from it the following bound:

Theorem D. *Let $w \in L^1(a,b)$ be a positive weight, and let u be a nontrivial solution of problem (1.1). Then*

$$\frac{\pi\sqrt{2}}{4} \leq \lambda \int_a^b (t-a)w(t)dt. \tag{1.4}$$

Theorem D is closely related to Bargmann's bound [4] for the number of negative eigenvalues of Schrödinger operators, although their bound on the left-hand side is better:

Theorem 1.4. *Let u be a solution of the stationary radial wave equation*

$$-u'' + \frac{l(l+1)}{r^2}u + w(x)u = 0$$

on $(0,\infty)$, with $u(0) = 0$ and vanishing at ∞, and $l \geq 0$ is the angular moment. Suppose that w satisfies

$$\int_0^\infty x|w(x)|dx < \infty.$$

Then

$$2l+1 < \int_a^b x|w(x)|dx.$$

As we will show later, the left-hand side of inequality (1.4) can be easily replaced by 1. However, we are interested in a similar inequality in the quasilinear setting, and we have the following result:

Theorem D'. *Let λ_1 be the first eigenvalue of*

$$-(|u'|^{p-2}u')' = \lambda w(x)|u|^{p-2}u, \qquad x \in (a,b),$$

with Dirichlet boundary conditions $u(a) = u(b) = 0$, and w is a positive weight in $L^1(a,b)$. Then we have

$$\frac{1}{p}\left(\frac{\pi_p}{2}\right)^{p-1} \leq \lambda_1 \int_a^b (x-a)^{p-1}w(x)dx.$$

Theorems D and D' are proved with the aid of Theorem C, using integral comparison theorems (see [76, 97, 106]) instead of Green's functions.

Let us observe that in Theorems C and D, the length of the interval does not appear in the inequality, and it make sense to take the limit as $L \to \infty$ whenever $w^{\frac{1}{2}}(x)$ or $xw(x)$ belongs to $L^1(0,\infty)$. This is their main advantage over Lyapunov's inequality. In this way, Theorems D and D' enable us to analyze several eigenvalue problems on the half-line:

$$-\frac{\mathrm{d}^2 u}{\mathrm{d} x^2} = \lambda w(x) u, \qquad x \geq 0,$$

with different boundary conditions at zero and infinity.

The first result about the eigenvalues of those problems is due to Hille [64]. Since then, it has been rediscovered many times. Birman and Solomyak [6] found a connection with the spectrum of Schrödinger operators in quantum waveguides (see also [105]). In recent years, Naito [85] and Elias [44] extended the result to higher-order operators, and Kusano and Naito [72] considered p-Laplacian operators; see also Drabek and Kufner [40].

1.2 Organization of the Book

The rest of this book is devoted to a study of Theorems A–D, their extension to quasilinear equations, and applications to eigenvalue problems.

We will study quasilinear problems involving the p-Laplacian equation

$$-(|u'|^{p-2}u')' = \lambda w(x)|u|^{q-2}u,$$

where $1 < p, q < \infty$. For $p = q$, we have a homogeneous problem, and the eigenvalues are well defined (see [39, 53, 57]). Also, for $p = q = 2$, we recover problem (1.1).

We will also consider *resonant systems* involving p- and q-Laplacians,

$$\begin{cases} -(|u'|^{p-2}u')' = \lambda \alpha w(x)|u|^{\alpha-2}|v|^{\beta}u, \\ -(|v'|^{q-2}v')' = \lambda \beta w(x)|u|^{\alpha}|v|^{\beta-2}v; \end{cases}$$

see the survey of Boccardo and De Figueiredo [8] for a beautiful introduction to this subject. Here, the exponents α and β satisfy

$$\frac{\alpha}{p} + \frac{\beta}{q} = 1,$$

recovering some homogeneity when we rescale $(u, v) \to (t^{\frac{1}{p}}u, t^{\frac{1}{q}}v)$.

We will study also some extensions to higher dimensions, and to higher-order operators.

More general operators in Orlicz spaces will be considered too. Here, given a convex, even, and positive function $\varphi : R \to R^+$ satisfying $\varphi(0) = 0$, we have the following ordinary differential equation:

$$-(\varphi(u'))' = \lambda w(x)\varphi(u); \tag{2.5}$$

see, for example, [55, 56, 83, 109]. In this case, only partial results are known about bounds for the spectrum; see [35].

Finally, let us mention the surveys on Lyapunov-type inequalities due to Cheng [25] and to Brown and Hinton [14]. For quasilinear problems, see Chap. 5 in the book of Dosly and Rehak [39], and Chap. 5 in the book of Pachpatte [92].

Appendix A

Since Theorem A is well known, only a brief description will be given in this appendix, together with a brief survey of the eigenvalue problem for quasilinear operators.

In this appendix we also include some necessary inequalities that will be used in the rest of the book.

Chapter 2

The Lyapunov inequality in Theorem B has been proved in many different works, both in the linear and nonlinear contexts.

We begin this chapter with different proofs for the linear case: we will review the original proof of Borg [9], the proof of Patula [93] by direct integration, and that of Nehari, who used Green's functions. We will describe the extension of Das and Vatsala [33] to higher-order equations, and we shall show that their proof implies the result of Hartman and Wintner for locally integrable weights. We include a Lyapunov-type inequality of Harris and Kong that takes care of the negative and positive parts of the weight.

The second section is devoted to p-Laplacian equations. According to [39], Lyapunov's inequality was proved first by Elbert in [43]. Then it was rediscovered several times, by Li and Yeh [78], and Pachpatte [91, 92], and it was proved almost simultaneously later in [75, 94, 114]. A different proof was given in [97]. We show the relationship between Lyapunov's inequality and integral comparison theorems, and we also consider higher-order quasilinear problems and nonconstant coefficients, and we analyze the optimality of the constants appearing in the inequalities.

We close the chapter with applications to eigenvalue problems.

Chapter 3

In Chap. 3, we will study Theorem C. First, we give the proof of Cohn in [28], which is reminiscent of Borg's proof of Lyapunov's inequality, followed by Calogero's proof [17], which is similar and uses the ideas behind Prüfer transformation methods. We also present the original proof of Nehari, together with his approach in [89], and the extension to p-Laplacian problems in [23], unifying the different cases.

We believe that Cohn and Calogero's proofs can be extended to p-Laplacian equations with some effort, but they cannot be generalized to include higher-order operators. On the other hand, Nehari claims that his proof holds for higher-order linear operators, and we write a full proof following their ideas. However, since this proof uses Green's functions, it cannot be used for nonlinear problems.

We also prove an extension to different powers of the weight. We also study the optimality of the bound, and we derive lower bounds of higher eigenvalues, including nonmonotonic weights with finitely many changes of monotonicity.

Chapter 4

In Chap. 4 we prove Bargmann's bound and Theorems D and D'. The rest of the chapter is devoted to some generalizations and the study of linear and quasilinear singular eigenvalue problems, like the ones considered in [6, 44, 64, 69, 72, 84, 85, 96]. This kind of bound provides information on the number of negative eigenvalues of Schrödinger operators as in [17, 28, 29, 105].

Let us note that Theorem D (with the sharp constant on the left-hand side) was obtained also by Kwong [73], who used Riccati equation techniques. Then it was generalized by Harris [61], and Harris and Kong [62].

The asymptotic behavior of eigenvalues for problems on the half-line was obtained in a few cases; we can cite only [64, 84, 96, 105]. Moreover, three different methods were used in those works: the Prüfer transform was used by Hille; then Naimark and Solomyak considered the s-numbers of the inverse operator, and I have used a variational approach. Remarkably, all these methods need to impose the monotonicity of the weight.

Chapter 5

In the third part, we present Lyapunov-type inequalities for general φ-Laplacian operators as in [35], and also for resonant systems (see [36]). Today, this is an active area of research, and several works have appeared in recent years.

Then we analyze the case of higher dimensions, which is far from being complete. To our knowledge, few results have been obtained in this case. We can cite the pioneering works of Egorov and Kondriatev [42], the recent ones of Cañada, Montero, and Villegas [19, 22] in the linear case, and [37] for p-Laplacian problems.

What Have We Left Out?

There are several generalizations of Theorem B that we cannot cover here. The following nonexhaustive list shows some of them, together with references:

1. Ordinary differential equations with other terms, such as:

Fink and Mary [51]	$u'' + p(x)u' = q(x)u,$
Ha [59]	$u'' + w(x)u + \lambda u = 0.$
Hochstadt [65]	$u^{(n)} - p(x)u^{(n-1)} = q(x)u.$

2. Reid [100, 102] introduced some inequalities for Hamiltonian systems

$$-v' = -C(t)u + A^T(t)v$$
$$u' = A(t)u + B(t)v,$$

where A, B, and C are real $n \times n$ matrix functions in $L^\infty(a,b)$, B symmetric and positive definite. Recently, several papers have extended those works.

3. Lyapunov inequalities for difference equations; see Cheng [24] and [116] for higher-order differences,

$$\Delta u(n) + w(n)u(n+1) = 0, \qquad n \in Z \cap [a,b].$$

Here, the proofs are similar to those obtained using Green's functions, although the constant is different according to the parity of $\#\{Z \cap [a,b]\}$.

4. Lyapunov-type inequalities for functional differential equations; see [45, 46]:

$$u''(x) + p(x)|u(x)|^{\mu-1}u(x) + m(x)|u(g(x))|^{\nu-1}u(g(x)) = 0,$$

where p, m, $g : [a,b] \to R$ are continuous and nonnegative functions, g strictly increasing, and μ, $\nu > 0$.

5. The relationship with other inequalities, such as those of Opial and Wirtinger.

6. Krein-Feller strings, Laplacians in fractals and Borel sets, dynamic equations in time scales; see [58, 82, 98].

Chapter 2
Lyapunov's Inequality

Abstract In this chapter we give some proofs of Lyapunov' inequality, in both the linear and nonlinear contexts.

2.1 The Classical Inequality

In this section we give some proofs of the classical Lyapunov's inequality in Theorem B. We consider the linear problem

$$-u'' = w(x)u, \qquad u(a) = u(b) = 0, \tag{1.1}$$

and we give different proofs of the inequality

$$\frac{4}{b-a} \le \int_a^b w(x)\,dx. \tag{1.2}$$

Throughout this chapter, we denote by c a point in (a,b) where $|u|$ is maximized.

2.1.1 The Linear Case

2.1.1.1 Borg's Proof

The first known proof is due to Borg [10], who attributed the inequality to Arne Beurling; see p. 68 in [9].

Let us suppose that $u \in C^2(a,b)$ is a positive solution of problem (1.1). Now, since $u(x) < u(c)$,

$$\int_a^b |w(x)|\,dx = \int_a^b |u''u^{-1}|\,dx > u(c)^{-1} \max_{a \le x < y \le b} |u'(y) - u'(x)|.$$

J.P. Pinasco, *Lyapunov-type Inequalities: With Applications to Eigenvalue Problems*,
SpringerBriefs in Mathematics, DOI 10.1007/978-1-4614-8523-0_2,
© Juan Pablo Pinasco 2013

We can write $c = a + s = b - t$ with $s + t = b - a$, and by Rolle's theorem, we can choose x and y such that

$$u'(y) = -u(c)t^{-1},$$
$$u'(x) = u(c)s^{-1},$$

so we get from the arithmetic–geometric–harmonic mean inequality (see Appendix A),

$$\int_a^b |w(x)|\,dx > \frac{1}{t} + \frac{1}{s} = \frac{s+t}{st} \geq \frac{4}{b-a},$$

and the proof is finished.

Remark 2.1. Let us note that in this proof, the weight is allowed to change sign, and the inequality involves $|w(x)|$. Aurel Wintner in [113] improved the bound by considering only the positive part of w, defined as

$$w^+(x) = \max\{w(x), 0\},$$

and let us introduce the negative part of w,

$$w^-(x) = \max\{-w(x), 0\},$$

which will be needed below.

The basic idea is to change w to w^+, and the Sturm oscillation theory implies that there exists a solution of

$$-u'' = w^+(x)u,$$

with $u(a) = u(b') = 0$, for some $b' < b$. Now we have

$$\int_a^b w^+(x)\,dx \geq \int_a^{b'} w^+(x)\,dx > \frac{4}{b'-a} > \frac{4}{b-a}.$$

2.1.1.2 Direct Integration

This is one of the simplest proofs, due to Patula [93]. It is based on the following Lemma:

Lemma 2.1. *Let u be a positive solution of problem (1.1), and let c be a point where u is maximized. Then*

$$(i)\ \int_a^c w^+(x)\,dx > \frac{1}{c-a},$$

$$(ii)\ \int_c^b w^+(x)\,dx > \frac{1}{b-c}.$$

$$(iii)\ \int_a^b w^+(x)\,dx > \frac{b-a}{(b-c)(c-a)}.$$

Proof. Let us assume that $u \geq 0$ in (a,b). By integrating u'', we get

$$u'(t) - u'(c) = \int_c^t w^-(s)u(s)ds - \int_c^t w^+(s)u(s)ds.$$

We integrate again, and interchanging the order of integration in the right-hand side, since $u'(c) = 0$, we get

$$u(t) - u(c) = \int_c^t (t-s)w^-(s)u(s)ds - \int_c^t (t-s)w^+(s)u(s)ds.$$

We choose $t = b$, and $u(b) = 0$ implies

$$u(c) = -\int_c^b (b-s)w^-(s)u(s)ds + \int_c^b (b-s)w^+(s)u(s)ds,$$

and since $w^- > 0$, we deduce that

$$u(c) \leq \int_c^t (b-s)w^+(s)u(s)ds \leq (b-c)u(c)\int_c^b w^+(s)ds,$$

and we get part (i) of the lemma.

Part (ii) follows in the same way.

Finally, part (iii) follows from the sum of (i) and (ii), and the lemma is proved. □

Remark 2.2. Let us note that Lyapunov's inequality follows from part (iii) in this lemma and the arithmetic–geometric–harmonic mean inequality.

It is interesting to observe that J.H.E. Cohn obtained Lemma 2.1 in [30], although the proof is more complex than the previous one.

Finally, let us observe that the motivation behind Patula's proof is the location of zeros of solutions, and he derived some oscillation results. Also, he found some conditions that guarantee that the distance between consecutive zeros of an oscillating solution goes to zero or infinity; see [93] for details.

2.1.1.3 Green's Functions and Higher-Order Problems

It is difficult to find the origin of the proofs based on the use of Green's functions. Brown and Hinton [14] cite the proof of Nehari in [86]. There is a long list of generalizations of Lyapunov's inequality based on previous techniques, and we can cite among others the works of Das and Vatsalaa [33] and Reid [100–102]. The main advantage of this method is its applicability to higher-order differential equations (compared with repeated integration), although it depends on the explicit knowledge of the corresponding Green's function.

Briefly, Nehari started with the Green's function of the operator $L(u) = u''$ with zero Dirichlet boundary conditions in (a,b), which is

$$G(x,y) = \frac{(x-a)(b-y)}{b-a},$$

and he wrote

$$u(x) = \int_a^b G(x,y)w(y)u(y)dy.$$

Then by choosing $x = c$, where $|u(x)|$ is maximized, he obtained

$$1 \leq \max_{a \leq x,y \leq b} |G(x,y)| \int_a^b |w(y)|dy$$

after canceling out $|u(c)|$ on both sides.

The arithmetic–geometric–harmonic mean inequality as in Remark 1.1 gives

$$\max_{a \leq x,y \leq b} |G(x,y)| \leq \frac{b-a}{4}.$$

Indeed, Nehari was working with complex-valued solutions of differential equations, and he used a curve in the complex plane, although the proof holds in the real case.

Remark 2.3. On the other hand, since the Green's function of elliptic operators is unbounded the existence of Lyapunov-type inequalities seems unlikely except in dimension one. We will see in Chap. 5 that it is possible to find some related inequalities.

For higher-order problems, the following Lyapunov's inequality was proved in [33], and is the model for similar developments involving more terms in the equation or different boundary conditions:

Theorem 2.1. *Let u be a nontrivial solution in (a,b) of*

$$(-1)^n u^{(2n)} = w(x)u$$

satisfying

$$u^{(k)}(a) = u^{(k)}(b) = 0, \qquad 0 \leq k \leq n-1.$$

Then

$$\int_a^b w^+(x)dx > \frac{4^{2n-1}(2n-1)[(n-1)!]^2}{(b-a)^{2n-1}}. \tag{1.3}$$

Proof. In the proof of Theorem 2.1, a little trick enables us to avoid the question of the sign of G. Here, we will need the explicit expression for $G(x,y)$ given in Das and Vatsala [33],

$$\frac{(-1)^{n-1}}{(2n-1)!}\left(\frac{(x-a)(b-y)}{b-a}\right)^n \sum_{j=0}^{n-1} \binom{n-1+j}{j}\left(\frac{(b-x)(y-a)}{b-a}\right)^j (y-x)^{n-1-j}$$

whenever $x \leq y$, and $G(x,y) = G(y,x)$.

Now, if u is a nontrivial solution, then

$$\int_a^b w(x)u^2 dx \leq \int_a^b w^+(x)u^2 dx$$

and

$$1 = \frac{\int_a^b [u^{(n)}]^2 dx}{\int_a^b w(x)u^2 dx} \geq \frac{\int_a^b [u^{(n)}]^2 dx}{\int_a^b w^+(x)u^2 dx} = \lambda_1 > 0,$$

where λ_1 is the first eigenvalue of

$$(-1)^n u^{(2n)} = \lambda w^+(x)u$$

with the boundary conditions

$$u^{(k)}(a) = u^{(k)}(b) = 0, \qquad 0 \leq k \leq n-1.$$

So from Mercer's theorem (see, for example, [31]), since $\lambda_1 \leq 1$, we obtain

$$1 \leq \sum_{k=1}^\infty \frac{1}{\lambda_k} = \int_a^b |G(x,x)| w^+(x) dx.$$

Now, $G(x,x) = 0$ when $x = y$, except for $j = n-1$ in the sum above. Therefore,

$$|G(x,x)| = \left| \frac{(-1)^{n-1}}{(2n-1)!} \left(\frac{(x-a)(b-x)}{b-a} \right)^n \binom{2n-2}{n-1} \left(\frac{(b-x)(x-a)}{b-a} \right)^{n-1} \right|$$

$$= \frac{(2n-2)!}{(n-1)!(n-1)!(2n-1)!} \left(\frac{(x-a)(b-x)}{b-a} \right)^{2n-1}$$

$$= \frac{1}{[(n-1)!]^2(2n-1)} \left(\frac{(x-a)(b-x)}{b-a} \right)^{2n-1}$$

gives the bound

$$[(n-1)!]^2(2n-1)(b-a)^{2n-1} \leq \int_a^b [(x-a)(b-x)]^{2n-1} w^+(x) dx,$$

and the proof of Theorem 2.1 follows by the arithmetic–geometric–harmonic mean inequality,

$$(x-a)(b-x) \leq \left(\frac{b-a}{2} \right)^2.$$

$\square$

Let us mention that this result was published without proof by Levin. For second- and fourth-order equations, it was proved also by Reid [101] (see the remark at the end of his paper).

The main difficulty of this proof is to obtain the Green's function G.

Remark 2.4. Recently, some works have appeared devoted to higher-order equations and Lyapunov's inequalities. However, some of the authors seem to be unaware of this proof, and in fact, they obtained less-precise bounds, such as

$$\frac{2^{n+1}n!}{(b-a)^{2n-1}} \le \int_a^b |w(x)|dx,$$

and even polynomial bounds in n instead of exponential ones. Also, there is no mention of Mercer's theorem, so they deal only with positive weights or bounds involving $|w|$. See [102] for a detailed analysis of the relationship between trace formulas and Lyapunov's inequality.

On the other hand, sometimes such proofs hold for quasilinear problems like the ones we will study in the next section, where no Green's functions are available. However, we will show an easy trick which gives much better bounds, recovering the same constant in the linear case.

2.1.1.4 Hartman–Wintner Proof

The proof involving Green's functions gives an immediate proof of the following inequality obtained by Hartman and Wintner [63]:

Theorem 2.2. *Let u be a nontrivial solution of problem (1.1). Then*

$$\int_a^b w^+(x)(b-x)(x-a)dx > b-a.$$

Let us recall that in the previous proof, we obtained

$$[(n-1)!]^2(2n-1)(b-a)^{2n-1} \le \int_a^b [(x-a)(b-x)]^{2n-1}w^+(x)dx.$$

Now the proof of Theorem 2.2 follows easily by replacing $n = 1$ in this bound. The original proof involved computations similar to those of Borg [10].

2.1.2 An Interesting Extension

Among several variants of Lyapunov's and related inequalities, one of the most interesting is the following, which was proved by Harris and Kong in [62].

Theorem 2.3. *Let u be a nontrivial solution of*

$$-u'' = w(x)u,$$

where $w \in C[a,b]$ and $u'(c) = 0$, $u(b) = 0$, with $u(x) \neq 0$ for all $x \in [c,b)$. Then

$$1 \leq (b-c) \sup_{c \leq t \leq b} \left| \int_c^t w(x)dx \right|. \tag{1.4}$$

Also, if $u(a) = u'(c) = 0$ and $u(x) \neq 0$ for every $x \in (a,c]$, then

$$a < (c-a) \sup_{a \leq t \leq c} \left| \int_t^c w(x)dx \right|. \tag{1.5}$$

Inequality (1.4) improves dramatically the previous bounds for an indefinite weight w, because it involves both the positive and negative parts of w, not only w^+, as in the previous proofs.

Proof. We follow [62]. Given u, after multiplying by (-1) if necessary, we can assume that $u > 0$. We now define the auxiliary function

$$v(x) = -\frac{u'(x)}{u(x)},$$

and a simple computation shows that v satisfies the first-order Riccati differential equation in (c,b):

$$v' = w(x) + v^2.$$

Clearly,

$$v(c) = 0, \qquad \lim_{x \to b^-} v(x) = \lim_{x \to b^-} -\frac{u'(x)}{u(x)} = +\infty,$$

and by integrating the equation from c to x, we get

$$v(x) = \int_c^x w(t)dt + \int_c^x v(t)^2 dt. \tag{1.6}$$

We now introduce the function

$$V(x) = \int_c^x v^2(t)dt.$$

Again, we have

$$V(c) = 0, \qquad \lim_{x \to b^-} V(x) = +\infty,$$

since

$$\lim_{x \to b^-} V(x) = \lim_{x \to b^-} \int_c^x v^2(t)dt = \lim_{x \to b^-} \left(v(x) - \int_c^x w(t)dt \right).$$

We obtain now a differential inequality that proves the theorem. To this end, let us define

$$W = \sup_{c \leq x \leq b} \left| \int_c^x w(s)ds \right|,$$

and let us note that Eq. (1.6) implies

$$V'(x) = v^2(x) \leq [W + V(x)]^2,$$

equivalently,

$$\frac{V'(x)}{[W + V(x)]^2} \leq 1.$$

Finally, by integrating this inequality in (c,d), we obtain

$$-\lim_{x \to b^-} \frac{1}{[W + V(x)]} + \frac{1}{W} \leq b - c,$$

which can be rewritten as

$$1 \leq (b - c) \sup_{c \leq x \leq b} \int_c^x w(s) ds.$$

Finally, observe that the inequality is strict. For if not, we must have

$$W(t) = \left| \int_c^t w(s) ds \right|$$

in (c,b). However, $W(t)$ is a continuous function, and $W(c) = 0$.

The other inequality follows in much the same way. $\square$

Remark 2.5. The inequality was extended to p-Laplacian operators in [66]. The proof is almost the same, although the inequalities now read

$$1 \leq (b - c)^{p-1} \sup_{c \leq x \leq b} \int_c^x w(s) ds,$$

$$1 \leq (c - a)^{p-1} \sup_{a \leq x \leq c} \int_x^c w(s) ds.$$

Recently, we obtained a different version of this inequality, which is valid for higher-order problems, when we are dealing with the asymptotic behavior of eigenvalues in homogenization problems; see [49]. It is possible to obtain from this inequality, with a different proof, a lower bound for the nth Dirichlet eigenvalue of a Sturm–Liouville problem. We present the proof for the linear case at the end of this chapter, together with some eigenvalue bounds.

2.2 Quasilinear Problems

We consider now the p-Laplacian problem in (a,b)

$$-(|u'|^{p-2})' = w(x)|u|^{p-2}u, \qquad x \in (a,b), \qquad (2.7)$$

with zero Dirichlet boundary conditions

$$u(a) = u(b) = 0, \tag{2.8}$$

and we prove the following Lyapunov's inequality for positive weights $w \in L^1(a,b)$:

$$\frac{2^p}{(b-a)^{p-1}} \leq \int_a^b w(x)\mathrm{d}x. \tag{2.9}$$

Clearly, for $p = 2$ we recover the classical Lyapunov's inequality.

We give two proofs below: the first was obtained by several authors, as we mentioned in the introduction. The second one shows the equivalence between Lyapunov's inequality and some integral comparison theorems. Moreover, we show the optimality of the constants involved in the bound by relating the problem to the Steklov eigenvalue problem, where the eigenvalue parameter appears in the boundary condition.

2.2.1 A Simple Proof

The following proof uses only the fact that $u \in W_0^{1,p}(a,b)$ can be written as

$$u(x) = \int_a^x u'(t)\mathrm{d}t,$$

together with Hölder's inequality and some elementary analysis.

We need also that

$$\int_a^b |u'|^p \mathrm{d}x = \int_a^b w(x)|u|^p \mathrm{d}x,$$

which is obtained by multiplying Eq. (2.7) by u and then integrating by parts.

Let $u \in W_0^{1,p}(a,b)$ be a nontrivial solution of problem (2.7) and (2.8), and let c be the point where $|u|$ is maximized. Then

$$2|u(c)| = \left| \int_a^c u'(x)\mathrm{d}x \right| + \left| \int_c^b u'(x)\mathrm{d}x \right|$$

$$\leq \int_a^b |u'(x)|\mathrm{d}x$$

$$\leq (b-a)^{\frac{p-1}{p}} \left(\int_a^b |u'(x)|^p \mathrm{d}x \right)^{\frac{1}{p}}$$

$$\leq (b-a)^{\frac{p-1}{p}} \left(\int_a^b w(x)|u|^p \mathrm{d}x \right)^{\frac{1}{p}}$$

$$\leq (b-a)^{\frac{p-1}{p}} \left(\int_a^b w(x)\mathrm{d}x \right)^{\frac{1}{p}} |u(c)|,$$

and by canceling out $|u(c)|$ and raising both sides to the power p, we get

$$2^p \le (b-a)^{p-1} \int_a^b w(x)dx,$$

which implies Lyapunov's inequality (2.9).

Remark 2.6. Clearly, when w is allowed to change sign, we have

$$\int_a^b w(x)|u|^p dx \le \int_a^b w^+(x)|u|^p dx,$$

and we obtain

$$\frac{2^p}{(b-a)^{p-1}} \le \int_a^b w^+(x)dx.$$

Remark 2.7. It is possible to obtain Lyapunov's inequality from a formulation similar that in Lemma 2.1. For p-Laplacian problems, we have

(i) $\displaystyle\int_a^c w^+(x)dx > \frac{1}{(c-a)^{p-1}}.$

(ii) $\displaystyle\int_c^b w^+(x)dx > \frac{1}{(b-c)^{p-1}}.$

(iii) $\displaystyle\int_a^b w^+(x)dx > \frac{(b-c)^{p-1}+(c-a)^{p-1}}{(b-c)^{p-1}(c-a)^{p-1}}.$

Instead of direct integration as in Patula's proof, we use the same argument as before,

$$\begin{aligned}
|u(c)|^p &= \left| \int_a^c u'(x)dx \right| \\
&\le (c-a)^{p-1} \int_a^c |u'(x)|^p dx \\
&\le (c-a)^{p-1} \int_a^b w^+(x)|u|^p dx \\
&\le (c-a)^{p-1}|u(c)| \int_a^b w(x)dx,
\end{aligned}$$

and (i) is proved. Now (ii) is identical, and (iii) follows from the sum of (i) and (ii).

2.2.2 Relationship with Integral Comparison Theorems

Here we prove a comparison theorem involving integral inequalities instead of pointwise ones, introduced in the linear case by Nehari [87]; see also St Mary [106], Levin [76]. The book of Dosly and Rehak [39] contains extensions to p-Laplacian equations. The proofs of Levin were based on Riccati-type equations, and Nehari used variational arguments. On the other hand, St Mary found a connection with Lyapunov-type inequalities, although he considered only monotonic functions. Our proof follows the work [97].

Theorem 2.4. *Let λ_1 and μ_1 be the first eigenvalues of problems*

$$-(|u'|^{p-2}u')' = \lambda \rho(x)|u|^{p-2}u, \qquad u'(c) = u(b) = 0, \tag{2.10}$$

$$-(|v'|^{p-2}v')' = \mu r(x)|v|^{p-2}v, \qquad v'(c) = v(b) = 0, \tag{2.11}$$

where $\rho, r \in C([a,b])$ are positive functions satisfying

$$\int_c^x r(t)dt \geq \int_c^x \rho(t)dt \tag{2.12}$$

for every $t \in [c,b]$. Then $\lambda_1 \geq \mu_1$, and equality is excluded unless $r(t) = \rho(t)$.

Proof. Let (λ_1, u) be an eigenpair of Eq. (2.10). We can take $u \geq 0$, and then $u' < 0$ in (c,b); if not, and $u'(c') = 0$ for some $c' \in (c,b)$, then

$$0 = |u'(c')|^{p-2}u'(c')| - |u'(c)|^{p-2}u'(c)| = \lambda_1 \int_c^{c'} \rho(x)|u|^{p-2}u(x)dx,$$

a contradiction to the fact that u and ρ are positive.

Starting from the weak formulation of the problem, we get

$$\begin{aligned}
\int_c^b |u'(x)|^p dx &= \lambda_1 \int_c^b \rho(x)|u(x)|^p dx \\
&= \lambda_1 \int_c^b \left(\int_c^x \rho(t)dt \right)' |u(x)|^p dx \\
&= \lambda_1 \left(\int_c^x \rho(t)dt \right) u(t)^p \Big|_c^b - \lambda_1 \int_c^b \left(\int_c^b \rho(t)dt \right) |u(x)^p|' dx \\
&= -\lambda_1 \int_c^b \left(\int_c^x \rho(t)dt \right) |u(x)^p|' dx.
\end{aligned}$$

From $u' < 0$, $u \geq 0$, we have $|u(x)^p|' = |u(x)|^{p-2}u(x)u'(x) < 0$, and then

$$-\int_c^b \left(\int_c^x \rho(t)dt \right) |u(x)^p|' dx \leq -\int_c^b \left(\int_c^x r(t)dt \right) |u(x)^p|' dx,$$

and integrating by parts again, we recover

$$-\int_c^b \left(\int_c^x r(t)dt \right) |u(x)^p|' dx = \int_c^b r(x)|u(x)|^p dx.$$

That is,

$$\int_c^b |u'(x)|^p dx \leq \lambda_1 \int_c^b r(x)|u(x)|^p dx.$$

Finally,

$$\mu_1 = \inf_{\{v \in W^{1,p}(c,b) : v(b)=0\}} \frac{\int_c^b |v'(x)|^p \mathrm{d}x}{\int_c^b r(x)|v(x)|^p \mathrm{d}x} \le \frac{\int_c^b |u'(x)|^p \mathrm{d}x}{\int_c^b r(x)|u(x)|^p \mathrm{d}x} \le \lambda_1,$$

and the Theorem is proved.   □

Remark 2.8. We can derive now Lyapunov's inequality by studying the following extremal problem, in the same spirit of Theorem 2.9, as we will see later:

Find a weight function $w \ge 0$ with $\int_c^b w(x)\mathrm{d}x = 1$ that minimizes the first eigenvalue of the problem

$$-(|u'|^{p-2}u')' = \lambda \rho(x)|u|^{p-2}u, \qquad u'(c) = u(b) = 0.$$

It is clear from the previous theorem that given any weight w, we can shift its mass slightly to the left, lowering the eigenvalue. So the optimal weight is not a function, but a Dirac delta distribution concentrated at c, i.e.,

$$\int_s^t \delta_c(x)f(x)\mathrm{d}x = \begin{cases} f(c) & c \in [s,t] \\ 0 & c \notin [s,t]. \end{cases}$$

In this limiting case, the Rayleigh quotient for the first eigenvalue reads

$$\lambda_1 = \inf_{\{v \in W^{1,p}(c,b) : v(b)=0\}} \frac{\int_c^b |v'(x)|^p \mathrm{d}x}{|v(c)|^p},$$

which coincides with the Rayleigh quotient of the one-dimensional Steklov eigenvalue problem,

$$\begin{cases} -(|u'|^{p-2}u')' = 0, \\ u(b) = 0, \\ |u'|^{p-2}u'(c) = \lambda|u(c)|^{p-2}u(c), \end{cases}$$

where the eigenvalue parameter appears in the boundary condition.

This problem has a unique eigenvalue, and can be computed explicitly: the eigenfunction is $u(x) = a(b-x)$, and we get at c,

$$|a|^{p-2}a = \lambda|a(b-c)|^{p-2}a(b-c),$$

that is,

$$\lambda = \frac{1}{(b-c)^{p-1}}.$$

This is equivalent to (ii) in Remark 2.7, since the equation

$$-(|u'|^{p-2}u')' = w(x)|u|^{p-2}u, \qquad u'(c) = u(b) = 0,$$

has a nontrivial solution if and only if $\lambda = 1$ is an eigenvalue of

$$-(|u'|^{p-2}u')' = \lambda w(x)|u|^{p-2}u$$

with the same boundary conditions. Thus, since $\lambda_1 \leq \lambda = 1$,

$$\frac{1}{(b-c)^{p-1}} \leq 1.$$

Let us remember that we are assuming $\int_c^b w(x)dx = 1$, and the general case follows by rescaling.

2.3 Some Incomplete Generalizations

Let us consider briefly here some problems that have been partially studied. We mention among them a Lyapunov-type inequality for higher-order operators without using Green's functions, the case of nonconstant-coefficient operators, and the optimality of the constants in those Lyapunov's inequalities.

2.3.1 Higher-Order Quasilinear Problems

There are few works devoted to higher-order quasilinear problems, mainly because the linear case was solved using Green's functions, which are not available now. Recently, Watanabe et al. [111] considered the embedding of the Sobolev spaces $H_0^m(a,b)$ into L^∞, and they computed the best constant C in the Sobolev inequality

$$\left(\sup_{x \in (a,b)} |u(x)| \right)^2 \leq C \int_a^b |u^{(m)}(x)|^2 dx$$

in terms of the Green's function of the operator $(-1)^m u^{2m}$.

We give here a simple generalization of inequality (1.3) for p-Laplacian problems of order $2m$, and we will use Taylor expansions instead of Green's functions.

Lemma 2.2. Let $1 < p < \infty$, and let $u \in C^{(m)}(a,b)$ satisfying

$$u^j(a) = 0, \qquad 0 \leq j \leq m-1.$$

Then

$$|u(x)|^p \le \left(\frac{p-1}{mp-1}\right)^{p-1} \frac{(x-a)^{mp-1}}{[(m-1)!]^p} \int_a^x |u^{(m)}(t)|^p dt.$$

Proof. Using Taylor's theorem and the Cauchy form of the remainder, we can write

$$u(x) = \sum_{j=0}^{m-1} \frac{u^j(a)}{j!}(x-a)^j + \int_a^x \frac{(x-t)^{m-1}}{(m-1)!} u^{(m)}(t) dt.$$

However, since the derivatives of u at a are zero, we have

$$u(x) = \int_a^x \frac{(x-t)^{m-1}}{(m-1)!} u^m(t) dt.$$

Thus by Hölder's inequality,

$$|u(x)| \le \left(\frac{p-1}{mp-1}\right)^{\frac{p-1}{p}} (x-a)^{\frac{mp-1}{p-1} \cdot \frac{p-1}{p}} \left(\int_a^x \left|\frac{u^{(m)}(t)}{(m-1)!}\right|^p dt\right)^{\frac{1}{p}}$$

$$\le (x-a)^{\frac{mp-1}{p}} \left(\frac{p-1}{mp-1}\right)^{\frac{p-1}{p}} \left(\int_a^x \left|\frac{u^{(m)}(t)}{(m-1)!}\right|^p dt\right)^{\frac{1}{p}},$$

and the proof is finished.   □

The following Lyapunov-type inequality is now a simple consequence of the previous Lemma:

Theorem 2.5. *Let $1 < p < \infty$, and let u be a nontrivial solution of*

$$-(|u^{(m)}|^{p-2} u^{(m)})^{(m)} = w(x)|u|^{p-2} u,$$

with Dirichlet boundary conditions

$$u^j(a) = u^j(b) = 0, \qquad 0 \le j \le m-1.$$

Then

$$\left(\frac{mp-1}{p-1}\right)^{p-1} \frac{2^{mp}[(m-1)!]^p}{(b-a)^{mp-1}} \le \int_a^b w(x) dx.$$

Proof. As usual, we choose c as the point where $|u|$ reaches its maximum. Now from Lemma 2.2,

$$\frac{|u(c)|^p}{(c-a)^{mp-1}} \le \left(\frac{p-1}{mp-1}\right)^{p-1} \frac{1}{[(m-1)!]^p} \int_a^c |u^{(m)}(t)|^p dt.$$

Also, starting from b, we have

$$\frac{|u(c)|^p}{(b-c)^{mp-1}} \le \left(\frac{p-1}{mp-1}\right)^{p-1} \frac{1}{[(m-1)!]^p} \int_c^b |u^{(m)}(t)|^p dt.$$

Hence, by adding both inequalities, we get

$$\frac{|u(c)|^p}{(c-a)^{mp-1}} + \frac{|u(c)|^p}{(b-c)^{mp-1}} \le \left(\frac{p-1}{mp-1}\right)^{p-1} \frac{1}{[(m-1)!]^p} \int_a^b |u^{(m)}(t)|^p dt,$$

and the left-hand side is minimized when both terms are equal,

$$(c-a)^{mp-1} = (b-c)^{mp-1} = \left(\frac{b-a}{2}\right)^{mp-1},$$

which gives

$$\frac{2^{mp}|u(c)|^p}{(b-a)^{mp-1}} \le \left(\frac{p-1}{mp-1}\right)^{p-1} \frac{1}{[(m-1)!]^p} \int_c^b |u^{(m)}(t)|^p dt. \qquad (3.13)$$

Finally, since

$$\int_a^b |u^{(m)}(x)|^p dx = \int_a^b w(x)|u(x)|^p dx \le |u(c)|^p \int_a^b w(x)dx,$$

after replacing the left-hand side of this inequality by the right-hand side in Eq. (3.13), after rearranging we obtain

$$|u(c)|^p \left(\frac{mp-1}{p-1}\right)^{p-1} \frac{2^{mp}[(m-1)!]^p}{(b-a)^{mp-1}} \le |u(c)|^p \int_c^b w(x)dx,$$

and the proof is finished. $\qquad\qquad\qquad\qquad\qquad\qquad\qquad\qquad\qquad\qquad\square$

Remark 2.9. Let us observe that for every p and $m = 1$, we recover Lyapunov's inequality. However, for $m > 1$ and $p = 2$, we have only

$$\frac{2^{2m}(2m-1)[(m-1)!]^2}{(b-a)^{2m-1}} \le \int_a^b w(x)dx,$$

and inequality (1.3) has a factor 4^{2m} instead of 2^{2m}.

Remark 2.10. To our knowledge, this is the best known result for higher-order quasi-linear equations, although we believe that the right constant involved must be 4^{mp}.

At least Theorem 2.5 gives the correct dependence on the length of the interval, and Lemma 2.2 gives a simple bound for $u(x)$ in terms of the higher derivatives. We can obtain in this way some inequalities for problems in which Green's functions cannot be used.

Remark 2.11. Recently, this problem was considered by Watanabe in [110], who found the optimal constant in the Sobolev embedding

$$W_0^{m,p}(a,b) \hookrightarrow L^\infty(a,b)$$

for $m = 1, 2$, and 3.

2.3.2 Nonconstant Coefficients

It is possible to extend Lyapunov's inequality to the following problem in (a,b):

$$-(v(x)|u'|^{p-2}u')' = w(x)|u|^{p-2}u, \qquad u(a) = u(b) = 0, \qquad (3.14)$$

assuming that v is an integrable function bounded away from 0.

There are two ways to obtain a Lyapunov-type inequality for this problem. The first, which can be found in [75, 114], is similar to the proof in Sect. 2.2.1,

$$\begin{aligned}
2|u(c)| &= \int_a^b |u'(x)|dx \\
&\leq \int_a^b |v^{\frac{-1}{p}}(x)v^{\frac{1}{p}}(x)u'(x)|dx \\
&\leq \left(\int_a^b v^{-\frac{1}{p-1}}\right)^{\frac{p-1}{p}} \left(\int_a^b v(x)|u'(x)|^p dx\right)^{\frac{1}{p}} \\
&\leq \left(\int_a^b v^{-\frac{1}{p-1}}\right)^{\frac{p-1}{p}} \left(\int_a^b w(x)|u|^p dx\right)^{\frac{1}{p}} \\
&\leq \left(\int_a^b v^{-\frac{1}{p-1}}\right)^{\frac{p-1}{p}} \left(\int_a^b w(x)dx\right)^{\frac{1}{p}} |u(c)|,
\end{aligned}$$

and we get the following result:

Theorem 2.6. *Let u be a nontrivial solution of problem (3.14), with $v, w \in L^1(a,b)$, and $v \geq c_1 > 0$. Then*

$$2^p \leq \left(\int_a^b v^{-\frac{1}{p-1}}\right)^{p-1} \int_a^b w(x)dx.$$

Remark 2.12. The other is to perform a change variables. In fact, if we define

$$P(x) = \int_a^x \frac{1}{v(s)^{\frac{1}{p-1}}}ds$$

and perform the change of variables

$$(x,u) \to (y,U),$$

where

$$y = P(x), \qquad U(y) = u(x),$$

after a simple computation we get

$$-(|\dot{U}|^{p-2}\dot{U})^{\cdot} = \lambda W(y)|U|^{p-2}U, \, y \in [0, L]$$
$$U(0) = U(L) = 0,$$

where

$$\cdot = d/dy,$$

$$L = \int_a^b \frac{1}{v(s)^{\frac{1}{p-1}}} ds,$$

and

$$W(y) = v(x)^{1/(p-1)} w(x).$$

Clearly, we can apply the Lyapunov inequality to this equation, obtaining

$$\frac{2^p}{L^p - 1} \le \int_0^L W(y) dy,$$

which is identical to the result in Theorem 2.6.

2.3.3 Singular Coefficients

An interesting problem is to find the right class of weights v and w admissible in this Lyapunov inequality, and the existence of other inequalities with different powers of the weights.

Remark 2.13. Recently, in [37], we considered A_p weights, the Muckenhoupt class of nonnegative functions $v \in L^1_{loc}(R^N)$ satisfying

$$\left(\int_B v(x) dx \right) \left(\int_B v(x)^{-\frac{1}{p-1}} dx \right)^{p-1} \le c_{p,v} |B|^p \qquad (3.15)$$

for every ball $B \in R^N$, with a fixed constant $c_{p,v}$ depending on the weight.

Moreover, we have the following result:

Theorem 2.7. *Let u be a nontrivial solution of problem (3.14), with $w \in L^1(a, b)$ and $v^p \in A_p$. Then*

$$\left(\frac{2}{b-a} \right)^p \le c_{p,v} \frac{\int_a^b w(x) dx}{\int_a^b v(x) dx}.$$

Let us note that for $v \equiv 1$, since $c_{p,1} = 1$, we recover again Lyapunov's inequality.

Remark 2.14. Theorem 2.7 extends the results in [75, 114], since it includes a large class of functions. Indeed, the A_p weights are exactly those for which the maximal operator of Hardy–Littlewood is bounded from L_v^p to L_v^p. When $1 < p < \infty$, the power functions $v(x) = |x|^\gamma$ belong to $A_p(R^N)$ for

$$-N < \gamma < N(p-1),$$

and hence we have a Lyapunov-type inequality for both degenerate and singular operators,

$$-(|x-x_0|^\gamma |u'|^{p-2}u')' = w(x)|u|^{p-2}u,$$

depending on the sign of γ, for every $x_0 \in (a,b)$.

We have analyzed in detail the case of $v(x) = d_{\partial\Omega}(x)$, the distance to the boundary for $\Omega \in R^N$ in [37]. Let us recall that functions of this kind (distance to a point or to the boundary) appear in several well-known problems, including Henon-type equations and Caffarelli–Kohn–Nirenberg inequalities.

2.3.4 Optimality of the Constants

The proof in Sect. 2.2.1 suggests that inequality (2.9) must be strict. If not, we must have in the last inequality

$$\int_a^b w(x)|u|^p \mathrm{d}x = |u(c)|^p \int_a^b w(x)\mathrm{d}x.$$

However, since u cannot be constant everywhere, w must be zero outside the points where $u(x) < u(c)$. Since the maximum of u must be unique between two consecutive zeros, u cannot be constant in any given interval. Thus, in order to get an equality, we need to have $w = \delta_c(x)$, the Dirac delta concentrated at c, and this is the solution of the extremal problem we have studied in the previous subsection. We have the following result, which is a corollary of the previous discussion:

Theorem 2.8. *The constant 2^p in Lyapunov's inequality (2.9) cannot be improved.*

Indeed, a similar argument can be applied in the proof involving Green's functions: from

$$[(n-1)!]^2(2n-1)(b-a)^{2n-1} \le \int_a^b [(x-a)(b-x)]^{2n-1}w^+(x)\mathrm{d}x,$$

and the arithmetic–geometric inequality, we observe that w must be concentrated on $x = (a+b)/2$. However, in both cases, $w \notin L^1(a,b)$.

Nevertheless, it makes sense to consider ordinary differential equations with measures on the right-hand side, since the functions in $W^{1,p}$ are continuous. We avoid this subject here, although it is related to several interesting problems that are currently being studied, such as differential equations in fractal sets, Stieltjes and Krein strings, and Sturm–Liouville operators with distributional coefficients.

Remark 2.15. Although Lyapunov's inequalities can be obtained for those problems, it is not clear that the constants are the same. Of course, higher-order problems are difficult even in the linear case if nonconstant coefficients are involved.

2.3.4.1 Optimality of the Power

Since the constant is optimal, different questions arise related to the norm of w: can we take a power of w other than 1, obtaining a similar inequality? Indeed, this question was solved by Egorov and Kondratiev for linear problems; see [42]. We summarize here their results.

Given $\beta \neq 0$, we denote by R_β the class of nonnegative measurable functions in $(0,1)$ such that

$$\int_0^1 w^\beta(x)dx = 1.$$

Let us define m_β as

$$m_\beta = \inf_{w \in R_\beta} \lambda_w,$$

where λ_w is the first eigenvalue of

$$-u'' = \lambda w(x)u, \qquad u(0) = u(1) = 0.$$

Egorov and Kondratiev's main result is the following:

Theorem 2.9. *Let R_β, m_β be defined as before.*

1. Let $\beta > 1$. Then

$$m_\beta = \frac{(\beta-1)^{1+\frac{1}{\beta}}}{\beta(2\beta-1)^{\frac{1}{\beta}}} B^2\left(\frac{1}{2}, \frac{1}{2} - \frac{1}{2\beta}\right),$$

where B is the Euler beta function,

$$B(a,b) = \int_0^1 x^{a-1}(1-x)^{b-1}dx.$$

2. Let $\beta = 1$. Then $m_\beta = 4$.
3. Let $\beta < 1$. Then $m_\beta = 0$.

Clearly, part (2) is Lyapunov's inequality; and both (1) and (3) characterize the powers β for which there exists a bound for every measurable function with $w^\beta \in L^1$.

The proof of (3) by Egorov and Kondratiev follows by considering a weight similar to

$$w(x) = \begin{cases} \varepsilon^{-\frac{1}{\beta}} & \text{if } x \in [\frac{1}{2} - \varepsilon, \frac{1}{2}] \\ \\ 0 & \text{if } x \notin [\frac{1}{2} - \varepsilon, \frac{1}{2}]. \end{cases}$$

Now, using as test function in the Rayleigh quotient the function

$$
u(x) = \begin{cases}
4x & \text{if } x \in [0, \tfrac{1}{4}], \\[2mm]
1 & \text{if } x \in [\tfrac{1}{4}, \tfrac{3}{4}], \\[2mm]
4 - 4x & \text{if } x \in [\tfrac{3}{4}, 1],
\end{cases}
$$

we obtain

$$
\lambda_w = \inf_{u \in H_0^1(0,1)} \frac{\int_0^1 u'^2 \, dx}{\int_0^1 w(x) u^2 \, dx} \le \frac{8}{\int_{\frac{1}{2}-\varepsilon}^{\frac{1}{2}} \varepsilon^{-\frac{1}{\beta}} \, dx} = 8\varepsilon^{\frac{1-\beta}{\beta}},
$$

which cannot be bounded from below by a fixed positive constant.

The proof of part (1) is more difficult, and it is related to the definition of the generalized trigonometric functions $\sin_p$, $\cos_p$, and $\sin_{pq}$; see the appendix. We include it because the same idea can be used with slight modifications for p-Laplacian problems.

Let w be a positive weight, $\int_0^1 w^\beta(x) dx = 1$ with $\beta > 1$. In order to compute

$$
m_\beta = \inf_{w \in R_\beta} \lambda_w,
$$

where λ_w is the first eigenvalue of

$$
-u'' = \lambda w(x) u, \qquad u(0) = u(1) = 0,
$$

let us start with the following estimate by applying Hölder's inequality:

$$
\int_0^1 w(x) u^2 \, dx \le \left(\int_0^1 w^\beta(x) \, dx \right)^{\frac{1}{\beta}} \left(\int_0^1 u^{\frac{2\beta}{\beta-1}}(x) \, dx \right)^{\frac{\beta-1}{\beta}} = \left(\int_0^1 |u|^q(x) \, dx \right)^{\frac{2}{q}},
$$

where we have set $q = 2\beta/(\beta - 1)$.

Let us take a sequence of weights w_k and the eigenfunctions u_{w_k} corresponding to the first eigenvalue λ_{w_k} such that

$$
\lim_{k \to \infty} \lambda_{w_k} = m_\beta.
$$

We can choose normalized eigenfunctions in order to have

$$
\int_0^1 |u_{w_k}|^q = 1,
$$

and then

$$
\int_0^1 u_{w_k}'^2 \to m_\beta \qquad \text{as } k \to \infty.
$$

Clearly, this is a bounded sequence in $H_0^1(0,1)$, which has a weakly convergent subsequence, and after renaming it u_{w_k}, we have a weak limit $u \in H_0^1(0,1)$ satisfying

$$\int_0^1 |u|^q dx = 1, \qquad \int_0^1 u'^2 dx \le \liminf_{k \to \infty} \int_0^1 u_{w_k}'^2 = m_\beta.$$

Indeed, since

$$\int_0^1 |u|^{(q-2)\beta} dx = \int_0^1 |u|^{\frac{2\beta^2}{\beta-1} - 2\beta} dx = \int_0^1 |u|^{\frac{2\beta}{\beta-1}} dx = 1,$$

the weight $|u|^{q-2}$ belongs to R_β, and then

$$m_\beta \le \int_0^1 u'^2 dx,$$

which gives the equality. Therefore,

$$m_\beta = \inf_{u \in H_0^1} \frac{\int_0^1 u'^2 dx}{\left(\int_0^1 |u|^{q-2} u^2 dx \right)^{\frac{2}{q}}},$$

and u solves the corresponding Euler–Lagrange equation associated with this minimization problem,

$$-u'' = m_\beta |u|^{q-2} u. \tag{3.16}$$

We have

$$-\int_0^x u'(t) u''(t) dt = m_\beta \int_0^x u^{q-1}(t) u'(t) dt,$$

that is, since $u(0) = 0$,

$$u'^2(0) - u'^2(x) = \frac{2m_\beta}{q} u^q(x),$$

and integrating again from 0 to 1, using the values of the integrals of u'^2 and u^q, we get

$$u'^2(0) - m_\beta = \frac{2m_\beta}{q},$$

which gives the value of $u'(0)$.

This gives a first-order ordinary differential equation for u,

$$u'(x) = m_\beta^{\frac{1}{2}} \left(\frac{q+2}{q} - \frac{2}{q} u^q(x) \right)^{\frac{1}{2}}, \tag{3.17}$$

and integrating from 0 to x, we obtain

$$\frac{1}{m_\beta^{\frac{1}{2}}} \int_0^x \frac{u'(x)\mathrm{d}x}{\left(\frac{q+2}{q} - \frac{2}{q}u^q(x)\right)^{\frac{1}{2}}} = x,$$

and after the change of variables

$$s = u(x),$$

we get

$$\frac{1}{m_\beta^{\frac{1}{2}}} \int_0^{u(x)} \frac{\mathrm{d}s}{\left(\frac{q+2}{q} - \frac{2}{q}s^q\right)^{\frac{1}{2}}} = x.$$

Let us consider again the differential equation (3.16). We know that every solution has at least a maximum. Moreover, the minimization problem admits both u and $|u|$ as solutions, so the uniqueness of solutions implies that $u \geq 0$. Also, since $u'' = -m_\beta|u|^{q-2}u \leq 0$, every critical point of u must be a maximum. This shows that there exists a unique point where the maximum is reached. Since the symmetry of the problem implies that $u(x)$ and $u(1-x)$ are solutions and therefore they are the same function, the maximum of u is attained at the center of the interval.

Let us set $M = u(1/2)$, and then

$$\frac{1}{m_\beta^{\frac{1}{2}}} \int_0^M \frac{\mathrm{d}s}{\left(\frac{q+2}{q} - \frac{2}{q}s^q\right)^{\frac{1}{2}}} = \frac{1}{2}.$$

Now from Eq. (3.17), since $u'(1/2) = 0$, we have

$$0 = m_\beta^{\frac{1}{2}} \left(\frac{q+2}{q} - \frac{2}{q}M^q\right)^{\frac{1}{2}},$$

that is,

$$M = \left(\frac{q+2}{2}\right)^{\frac{1}{q}}.$$

Changing variables again, $s = Mt$, gives

$$\frac{1}{m_\beta^{\frac{1}{2}}} \int_0^1 \frac{M\mathrm{d}t}{\left(\frac{q+2}{q} - \frac{2}{q}(Mt)^q\right)^{\frac{1}{2}}} = \frac{1}{2},$$

and replacing M and rearranging terms yields

$$\frac{\left(\frac{q+2}{2}\right)^{\frac{1}{q}-\frac{1}{2}}}{\left(\frac{2m_\beta}{q}\right)^{\frac{1}{2}}}\int_0^1\frac{dt}{(1-t^q)^{\frac{1}{2}}}=\frac{1}{2}.$$

Let us recall the Euler beta function

$$B(a,b)=\int_0^1 x^{a-1}(1-x)^{b-1}dx.$$

Clearly, by changing variables again, $x=t^q$, we obtain

$$\int_0^1\frac{dt}{(1-t^q)^{\frac{1}{2}}}=\frac{1}{q}\int_0^1(1-x)^{-\frac{1}{2}}x^{\frac{q}{q-1}}dx=\frac{1}{q}B\left(\frac{1}{q},\frac{1}{2}\right).$$

So,

$$\frac{\left(\frac{q+2}{2}\right)^{\frac{1}{q}-\frac{1}{2}}}{q\left(\frac{2m_\beta}{q}\right)^{\frac{1}{2}}}B\left(\frac{1}{q},\frac{1}{2}\right)=\frac{1}{2},$$

and we obtain the constant m_β,

$$m_\beta=\frac{2}{q}\left(\frac{q+2}{2}\right)^{\frac{2}{q}-1}B^2\left(\frac{1}{q},\frac{1}{2}\right),$$

Finally, recalling that we have set $q=2\beta/(\beta-1)$ and replacing this in the previous formula gives

$$m_\beta=\frac{(\beta-1)^{1+\frac{1}{\beta}}}{\beta(2\beta-1)^{\frac{1}{\beta}}}B^2\left(\frac{1}{2}-\frac{1}{2\beta},\frac{1}{2}\right),$$

as desired.

Remark 2.16. Observe that the length of the interval is missing, since we are working in $[0,1]$. A simple change of variables gives the explicit power for each $\beta\geq 1$.

2.4 Eigenvalue Problems: Lower Bounds of Eigenvalues

Theorem 2.9 in the previous section gives a clear relationship between Lyapunov-type inequalities and eigenvalue problems. Let us note that changing w to λw, in both the linear and quasilinear cases, we easily obtain a lower bound for the first eigenvalue. Moreover, we can bound higher eigenvalues using the fact that u_n, the eigenfunction associated with the nth eigenvalue, has exactly $n+1$ zeros.

Theorem 2.10. *Let λ_n be the nth eigenvalue of*

$$-(|u'|^{p-2})' = \lambda w(x)|u|^{p-2}u, \qquad x \in (a,b), \tag{4.18}$$

with zero Dirichlet boundary conditions $u(a) = u(b) = 0$. Then

$$\frac{2^p n^p}{(b-a)^{p-1}\int_a^b w(x)dx} \leq \lambda_n. \tag{4.19}$$

Proof. Let $a = x_0 < x_1 < \cdots < x_n = b$ be the zeros of an eigenfunction u_n associated with λ_n. Using Lyapunov's inequality in the n intervals (x_{i-1}, x_i), we get

$$\sum_{k=1}^n \frac{2^p}{(x_k - x_{k-1})^{p-1}} \leq \lambda_n \sum_{k=1}^n \left(\int_{x_{k-1}}^{x_k} w(x)dx\right) \leq \lambda_n \int_a^b w(x)dx.$$

Since the sum on the left-hand side is minimized when all the terms are the same, we obtain the lower bound

$$2^p n \left(\frac{n}{b-a}\right)^{p-1} \leq \lambda_n \int_a^b w(x)dx,$$

and the proof is finished.   □

2.4.1 Optimality of the Bound

We have proved previously that Lyapunov's inequality is sharp, and we can ask about the optimality of the bounds in Theorem 2.10. We will follow the ideas in [94] to show that they are sharp, too, and that there exists some family of positive weights $\{w_n^j\}_j$ with $\int_a^b w_n^j(x)dx = 1$ such that $\lambda_n^j \to 2^p n^p (b-a)^{1-p}$.

Theorem 2.11. *There exists a family of weights functions $\{w_n^j\}_j$ with $\int_a^b w_n^j(x)dx = 1$ such that*

$$\lim_{j\to\infty} \lambda_n^j = \frac{2^p n^p}{(b-a)^{p-1}},$$

where λ_n^j is the nth eigenvalue of

$$-(|u'|^{p-2}u')' = \lambda w_n^j |u|^{p-2}u$$

with zero Dirichlet boundary conditions $u(a) = u(b) = 0$.

Proof. Let us consider the first eigenvalue λ_1, and let $c = (a+b)/2$.
Let $w_1 = \delta_c(x)$ be the Dirac delta function centered at c. We have

$$\lambda_1 = \min_{u\in W_0^{1,p}} \frac{\int_a^b |u'|^p}{\int_a^b \delta_c |u|^p} = \min_{u\in W^{1,p}:u(a)=0} 2\frac{\int_a^c |u'|^p}{|u(c)|^p}.$$

Let us observe that we have obtained the Rayleigh quotient for the first eigenvalue of the Steklov problem

$$\begin{cases} -(|u'(x)|^{p-2}u'(x))' = 0, \\ |u'(c)|^{p-2}u'(c) = \mu|u(c)|^{p-2}u(c), \\ u(a) = 0, \end{cases} \tag{4.20}$$

and therefore

$$\lambda_1 = 2\mu_1.$$

We can compute the solutions explicitly, which are constant multiples of $u(x) = x - a$. Hence,

$$\mu_1 = \frac{2^{p-1}}{(b-a)^{p-1}}.$$

Now we define the functions w_1^j,

$$w_1^j = \begin{cases} 0 & \text{for } x \in \left[a, \dfrac{a+b}{2} - \dfrac{1}{j}\right], \\[2mm] \dfrac{j}{2} & \text{for } x \in \left[\dfrac{a+b}{2} - \dfrac{1}{j}, \dfrac{a+b}{2} + \dfrac{1}{j}\right], \\[2mm] 0 & \text{for } x \in \left[\dfrac{a+b}{2} + \dfrac{1}{j}, b\right], \end{cases}$$

and the result follows by testing in the Rayleigh quotient the first Steklov eigenfunction,

$$u(x) = \begin{cases} x - a & \text{if } x \in \left[a, \dfrac{a+b}{2}\right], \\[2mm] b - x & \text{if } x \in \left[\dfrac{a+b}{2}, b\right], \end{cases}$$

since $w_1^j \rightharpoonup \delta_c$. Hence the inequality is sharp for $n = 1$.

The case $n \geq 2$ follows in much the same way, by dividing the interval $[a, b]$ into n subintervals I_i of equal length. Taking the midpoint c_i of each subinterval, we introduce the weight

$$w_n(x) = \frac{1}{n} \sum_{i=1}^{n} \delta_{c_i}(x),$$

and by symmetry (and uniqueness) of solutions, the restriction of u_n to I_i is the eigenfunction in this interval. So, since

$$\mu_1 = \frac{1}{\left(\frac{b-a}{n}\right)^{p-1}},$$

we get

$$\lambda_n = 2n\mu_1 = \frac{2^p n^p}{(b-a)^{p-1}},$$

and a similar approximation argument as before finishes the proof.   □

2.4.2 A Different Bound

The following result was obtained recently in [49] for higher-order ordinary differential equations and p-Laplacian problems.

We were interested in homogenization problems where the weight changes sign. For such problems, we have two sequences of eigenvalues tending to $\pm\infty$. However, the limit problem is formulated in terms of the mean value of the weight, so at least one of these sequences will disappear. A Lyapunov-type inequality enables us to prove this result.

The following theorem is representative of the kind of bounds that can be obtained, and for brevity, we consider here only the linear case. Observe that there are some similarities with Theorem 2.3, although the proof is based on variational arguments instead of Riccati equation techniques.

Theorem 2.12. *Let u be an eigenfunction associated to λ_n, the nth eigenvalue of*

$$-u'' = \lambda w(x)u, \qquad u(a) = u(b) = 0,$$

with $w \in L^1[a,b]$. Then

$$\frac{\pi n}{b-a} \leq 2\lambda_n \sup_{a \leq x \leq b} \left| \int_a^x w(t)\mathrm{d}t \right|. \tag{4.21}$$

Proof. By multiplying the equation by u and integrating by parts, we get

$$\int_a^b u'^2 \mathrm{d}x = \lambda_n \int_a^b w(x)u^2.$$

Let us introduce now the auxiliary function

$$W(x) = \int_a^x w(t)\mathrm{d}t,$$

and we can write

$$\int_a^b w(x)u^2 \mathrm{d}x = \int_a^b W'(x)u^2 \mathrm{d}x = -\int_a^b W(x)2uu' \mathrm{d}x.$$

Since

$$\left| \int_a^b W(x)2uu' \mathrm{d}x \right| \leq 2 \sup_{a \leq x \leq b} |W| \int_a^b |u||u'| \mathrm{d}x,$$

Hölder's inequality implies that

$$\left(\int_a^b u'^2 dx \right)^2 \leq 4\lambda_n^2 \sup_{a \leq x \leq b} |W|^2 \int_a^b u^2 dx \int_a^b u'^2,$$

and after canceling,

$$\frac{\int_a^b u'^2 dx}{\int_a^b u^2 dx} \leq 4\lambda_n^2 \sup_{a \leq x \leq b} |W|^2.$$

For $n = 1$, we have

$$\frac{\pi}{b - a} \leq 2\lambda_1 \sup_{a \leq x \leq b} |W|, \tag{4.22}$$

since u is an admissible function in the variational characterization of the first eigen-value for the constant-coefficients problem, and

$$\frac{\pi^2}{(b - a)^2} \leq \frac{\int_a^b u'^2 dx}{\int_a^b u^2 dx}.$$

For higher eigenvalues, let us observe that u has n nodal domains $\{I_i\}_{i=1}^n$, and we can take n functions $\{u_1, \dots, u_n\}$ by restricting u at each nodal domain. Each u_i is a solution of the weighted problem in the corresponding interval I_i, and thus, applying the previous inequality, we have

$$\frac{\pi^2}{|I_i|^2} \leq 4\lambda_1(I_i)^2 \sup_{x \in I_i} |W|^2 \leq 4\lambda_n^2 \sup_{a \leq x \leq b} |W|^2,$$

since $\lambda_1(I_i) = \lambda_n$. Hence

$$\frac{1}{n} \sum_{i=1}^n \frac{\pi^2}{|I_i|^2} \leq 4\lambda_n^2 \sup_{a \leq x \leq b} |W|^2,$$

and the left-hand side is minimized when all the lengths $|I_i|$ are equal, which gives

$$\frac{\pi^2 n^2}{(b - a)^2} \leq 4\lambda_n^2 \sup_{a \leq x \leq b} |W|^2,$$

and the proof is finished.   □

Remark 2.17. The proof is almost identical for second-order quasilinear equations. However, for higher-order equations (linear or not), we cannot use the argument of nodal domains, since a Dirichlet eigenfunction has a multiple zero at the boundary of the interval and the interior zeros are simple. In this case, the proof follows as in Theorem 2.5.

Chapter 3
Nehari–Calogero–Cohn Inequality

Abstract In this chapter, we review the proofs of Nehari, Calogero, and Cohn for Theorem C, together with some generalizations for the p-Laplacian eigenvalues and higher-order problems.

> The motivation for a physicist to study 1-dimensional problems
> is best illustrated by the story of the man who,
> returning home late at night after an alcoholic evening,
> was scanning the ground for his key under a lamppost;
> he knew, to be sure, that he had dropped it somewhere else,
> but only under the lamppost was there enough light to conduct a proper search.
>
> —F. Calogero

3.1 The Work of Calogero and Cohn

In this section we prove Theorem C', and we review the proofs of Cohn and Calogero. Let us recall that the problem under consideration is

$$-u'' = w(x)u, \qquad u(a) = u(b) = 0, \tag{1.1}$$

where w is a monotonic weight, and we show that a necessary condition for the existence of a nontrivial solution is

$$\frac{\pi}{2} \le \int_a^b |w(x)|^{\frac{1}{2}} dx. \tag{1.2}$$

We also include a generalization of Theorem C' due to Cohn under concavity hypotheses on w.

J.P. Pinasco, *Lyapunov-type Inequalities: With Applications to Eigenvalue Problems*,
SpringerBriefs in Mathematics, DOI 10.1007/978-1-4614-8523-0_3,
© Juan Pablo Pinasco 2013

Before the proofs, let us observe that by Hölder's inequality, we get

$$\frac{\pi^2}{4} \le \left(\int_a^b |w(x)|^{\frac{1}{2}} dx \right)^2 \le (b-a) \int_a^b w(x) dx,$$

which is a Lyapunov-type inequality, at least for monotonic weights, with a different—and worse—constant, since $\pi^2/4 < 4$, although with the correct homogeneity on the length of the interval and the weight. Of course, similar inequalities can be derived with different norms of w.

3.1.1 Cohn's Proof

The proof of Cohn in [28] is very detailed, and all the assumptions are carefully analyzed, such as the case of increasing or decreasing weights, the fact that a and b can be considered two consecutive zeros, and the uniqueness of c with $u'(c) = 0$ for consecutive zeros. Cohn also considered sign-changing weights, and he proved the optimality of the bound. For the sake of completeness, we reproduce here Cohn's arguments.

Lemma 3.1. *Let u be a nontrivial solution of*

$$u'' + w(x)u = 0, \qquad u(a) = u(b) = 0,$$

with a monotonic weight w. In order to prove Theorem C', we can replace w by w^+, and we can assume that w is decreasing and that a, b are consecutive zeros. Moreover, in this case, there exists a single point c where $u' = 0$.

Proof. Let us note first that if w changes sign, then w^+ is also monotonic. Now, since $w^+ \ge w$, the Sturmian theory (see Theorem 1.3) shows that every solution of the problem

$$v'' + w^+(x)v = 0, \qquad v(a) = 0,$$

has another zero $b' \in (a, b)$. So,

$$\int_a^b |w(x)|^{\frac{1}{2}} dx \ge \int_a^b w^+(x)^{\frac{1}{2}} dx \ge \int_a^{b'} w^+(x)^{\frac{1}{2}} dx.$$

The same argument enables us to consider a, b' two consecutive zeros of u.

For increasing weights w, we can change variables, and we obtain the problem

$$v'' + w(-x)v = 0, \qquad u(-b) = u(-a) = 0,$$

where the weight is now decreasing in $(-b, -a)$. Clearly,

$$\int_a^b w(x)^{\frac{1}{2}} dx = \int_{-b}^{-a} w(-x)^{\frac{1}{2}} dx.$$

For nonnegative and decreasing w, the existence of $c \in (a,b)$ with $u'(c) = 0$ follows from Rolle's theorem, since $u(a) = u(b) = 0$. Moreover, $w(c) > 0$, or $w \equiv 0$ in (c,b), and we get

$$0 = \int_c^b w(x)u\,dx = \int_c^b u''\,dx = u'(b) \neq 0,$$

a contradiction, since u and u' cannot both be zero at b or $u \equiv 0$. Hence, if there existed another zero d of u', we must have $w > 0$ in (c,d) (applying the same argument in (d,b)), and then

$$0 = \int_c^d u''\,dx = \int_c^d w(x)u\,dx > 0,$$

which is impossible.

The lemma is proved.   □

Let us now prove Theorem C' for a nonincreasing positive weight w. Let c be as given by the previous lemma, the unique zero of u', and let us introduce the function $\theta(x) : [a,c] \to [0,\pi/2]$ such that

$$u'\sin(\theta(x)) = u(x)w(x)^{\frac{1}{2}}\cos(\theta(x)).$$

Rewriting it as

$$\frac{u'(x)}{u(x)} = w(x)^{\frac{1}{2}}\cot(\theta(x)),$$

we obtain

$$\left(\frac{u'(x)}{u(x)} - \frac{u'(x)^2}{u(x)^2}\right)\,dx = d[w(x)^{\frac{1}{2}}\cot(\theta(x))].$$

A bit of trigonometry shows that the term inside the parentheses on the left-hand side is equal to $-w\csc^2(\theta) = -w\sin^{-4}(\theta)$. Then changing variables and using the signs of $\sin(\theta)$, $\cos(\theta)$ and the fact that q is a decreasing function, we get

$$\int_a^c w(x)^{\frac{1}{2}}\,dx = \int_{\theta(a)}^{\theta(c)} w^{-\frac{1}{2}}\sin^2(\theta)d[w^{\frac{1}{2}}\cot(\theta)]$$

$$= \int_{\theta(a)}^{\theta(c)} d\theta - w^{-\frac{1}{2}}\sin(\theta)\cos(\theta)d[w^{\frac{1}{2}}]$$

$$\geq \int_{\theta(a)}^{\theta(c)} d\theta$$

$$= \theta(c) - \theta(a)$$

$$= \frac{\pi}{2},$$

and the proof of Theorem C' is finished.

3.1.2 Calogero's Proof

Francesco Calogero in [17] states inequality (1.2) in physical terms and presents it as a condition for the existence of bound states for a one-dimensional Schrödinger equation in $(0,\infty)$. His proof is very similar to the previous one, although less structured. We modify it in order to unify the notation.

First, Calogero chooses the first point $c \in (a,b)$ with $u'(c) = 0$, and then he introduces the function $v(x)$ defined as

$$\tan(v(x)) = w(x)^{\frac{1}{2}} \frac{u(x)}{u'(x)}. \tag{1.3}$$

Observe the similarity between v and θ defined in the previous proof.

A straightforward computation shows that v satisfies the following first-order differential equation:

$$v'(x) = w(x)^{\frac{1}{2}} - \frac{1}{4} \frac{w'(x)}{w(x)} \sin(2v(x)),$$

with the initial condition $v(0) = 0$.

Now, v is increasing, since $v'(0) = w(0) > 0$, and $v(c) = \pi/2$. Indeed, $0 \le v \le \pi/2$ in (a,c), since c is the first zero of u'. Moreover, $\sin(2v(x))$ is nonnegative, together with u and u'. So after bounding

$$0 \le \sin(2v(x)) \le 1,$$

we get the differential inequality

$$v'(x) \le w(x)^{\frac{1}{2}}.$$

Trivially, we now obtain

$$\int_a^b w(x)^{\frac{1}{2}} dx \ge \int_a^c w(x)^{\frac{1}{2}} dx \ge \int_a^c v'(x) dx = v(c) - v(a) = \frac{\pi}{2},$$

and the inequality is proved.

Remark 3.1. Both Cohn and Calogero went on to derive upper and lower bounds for the number of negative eigenvalues of one-dimensional Schrödinger equations; see [18, 29].

Remark 3.2. We believe that those proofs can be extended to p-Laplacian problems, since we have a Sturmian theory, and that Riccati-type techniques can be used, although the computations seem to be harder. Later, we overcome the difficulties by constructing a different type of proof.

3.1.3 A Partial Converse

Cohn, in his second paper on this subject [29], obtained the following partial converse of Theorem C', under different conditions on the weight. The same result was obtained previously by Makai [81]:

Theorem 3.1. *Let* $w^{-\frac{1}{4}} \in C^2(a,b)$ *be a convex function. If u is a nontrivial solution of*

$$u'' + w(x)u = 0$$

with two consecutive zeros a and b, then

$$\int_a^b w(x)^{\frac{1}{2}}dx \leq \pi.$$

Proof. The proof of this theorem follows by a comparison argument. Cohn introduced the new variable

$$s(x) = \int_a^x w(t)^{\frac{1}{2}}dt.$$

Then the function

$$v(s) = w(x)^{\frac{1}{4}}u(x)$$

satisfies the following differential equation:

$$\frac{d^2v}{ds^2} + \left(1 + w^{\frac{3}{4}}\frac{d^2 w^{-\frac{1}{4}}}{dx^2}\right)v = 0.$$

From the convexity hypotheses in $w^{-\frac{1}{4}}$, the term inside the parentheses is greater than 1. So if the new variable satisfies

$$s(b) = \int_a^b w(x)^{\frac{1}{2}}dx > \pi,$$

we cannot have two consecutive zeros a, b of u, since v has a zero $b' < b$ as a consequence of the Sturmian comparison theorem, and both functions share the same zeros.

The proof is finished. □

Remark 3.3. There are several works devoted to similar upper bounds under different hypotheses on the weights. We omit them to shorten this presentation.

3.2 Nehari's Proof and Generalizations

Nehari's proof is very different from the previous proofs and is more abstract. The advantage is that we can apply it to higher-order equations, although it is restricted to linear operators.

For convenience, let us start with $\mathscr{L}$, a self-adjoint linear differential operator of order $2n$ and the following problem with zero Dirichlet boundary conditions:

$$\begin{cases} \mathscr{L}u(x) = f(x) & x \in (a,b), \\ y^{(i)}(a) = y^{(i)}(b) = 0 & 0 \le i \le n-1. \end{cases} \tag{2.4}$$

Here, the solution can be written as

$$u(x) = \int_a^b G(x,y)f(y)dy,$$

where G is the Green's function corresponding to the operator $\mathscr{L}$.

For the eigenvalue problem

$$\begin{cases} \mathscr{L}u(x) = \lambda w(x)u & x \in (a,b) \\ y^{(i)}(a) = y^{(i)}(b) = 0 & 0 \le i \le n-1, \end{cases} \tag{2.5}$$

we have the following result:

Theorem 3.2. *Let λ_1 be the first eigenvalue of problem (2.5). Then*

$$c(n) < \lambda_a^{\frac{1}{2n}} \int_a^b w^{\frac{1}{2n}}(x)dx,$$

where $c(n)$ depends only on the order of the equation. In particular, for $n = 1$,

$$c(1) = \frac{\pi}{2}.$$

3.2.1 Nehari's Proof for Second-Order Problems

3.2.1.1 Notation and Preliminary Results

Let us note that we can analyze the equivalent integral equation

$$\mu u(x) = \int_a^b G(x,y)w(y)u(y)dy,$$

where $\mu = \lambda^{-1}$.

Now multiplying both sides by $w^{\frac{1}{2}}(x)$ and introducing the new unknown function $v(x) = u(x)w^{\frac{1}{2}}(x)$, we transform the problem to

$$\mu v(x) = \int_a^b G(x,y)w^{\frac{1}{2}}(x)w^{\frac{1}{2}}(y)v(y)dy. \tag{2.6}$$

Let us recall that the first eigenvalue is obtained with the Rayleigh method,

$$\mu_1 = \sup_{\{u \in L^2 : \|u\|_2 = 1\}} \int_a^b \int_a^b G(x,y) w^{\frac{1}{2}}(x) w^{\frac{1}{2}}(y) v(x) v(y) dxdy. \qquad (2.7)$$

Of course, this is just a particular case, with $g = w^{\frac{1}{2}}$, $K(x,y) = G(x,y)$, of the general eigenvalue problem

$$\mu u(x) = \int_a^b K(x,y) g(x) g(y) u(y) dy. \qquad (2.8)$$

We will call μ_g the first eigenvalue and u_g the associated eigenfunction. Moreover, let us denote by $J(g,g;u)$ the double integral

$$J(g,g;u) = \int_a^b \int_a^b K(x,y) g(x) g(y) u(x) u(y) dxdy.$$

Let us note that we can write it as

$$J(g,g;u) = J(g^{\frac{1}{2}}(x) g^{\frac{1}{2}}(y), g^{\frac{1}{2}}(x) g^{\frac{1}{2}}(y); u(x) u(y)),$$

which implies, since $K \in L^2[(a,b) \times (a,b)]$, that

$$J(f,g;u) \le [J(f,f;u) J(g,g;u)]^{\frac{1}{2}}. \qquad (2.9)$$

In fact, Hölder's inequality implies

$$J(f,g;u) \le \int_a^b \int_a^b K(x,y) f^{\frac{1}{2}}(x) f^{\frac{1}{2}}(y) g^{\frac{1}{2}}(x) g^{\frac{1}{2}}(y) u(x) u(y) dxdy$$

$$\le \left(\int_a^b \int_a^b K(x,y) f(x) f(y) u(x) u(y) dxdy \right)^{\frac{1}{2}}$$

$$\cdot \left(\int_a^b \int_a^b K(x,y) g(x) g(y) u(x) u(y) dxdy \right)^{\frac{1}{2}}$$

$$= [J(f,f;u) \cdot J(g,g;u)]^{\frac{1}{2}}.$$

Moreover, we can write with this new notation,

$$\mu_g = J(g,g;u_g) \ge J(g,g;u) \qquad (2.10)$$

for every $u \in L^2$ with $\|u\|_2 = 1$. This implies that $u_g \ge 0$, since $g > 0$, $K > 0$, and

$$\mu_g = J(g,g;u_g) \le |J(g,g;u_g)| \le J(g,g;|u_g|),$$

since the eigenvalues of the one-dimensional differential problem are simple.

Finally, let us introduce the following sets of functions:

$$M = \{g : g \in L^2([a,b]) \text{ nonnegative and monotonic}\},$$

$$R = \{r : r \in L^2([a,b]) \text{ monotonic step function}\},$$

$$S = \{s : s \in R \text{ with at most a single discontinuity point in } [a,b]\}.$$

3.2.1.2 A Key Lemma

The following result is the key point in Nehari's proof:

Lemma 3.2. *Given the eigenvalue problem* (2.8),

$$\inf_{g \in M} \left\{ \mu_g^{-\frac{1}{2}} \int_a^b g(x) dx \right\} = \inf_{r \in R} \left\{ \mu_r^{-\frac{1}{2}} \int_a^b r(x) dx \right\} = \inf_{s \in S} \left\{ \mu_s^{-\frac{1}{2}} \int_a^b s(x) dx \right\}. \quad (2.11)$$

Proof. Given $g \in M$ and $\varepsilon > 0$, by density there exists some step function r such that

$$\int_a^b [g(x) - r(x)]^2 dx < \varepsilon^2.$$

Now, using Hölder's inequality,

$$\left| \int_a^b g(x) dx - \int_a^b r(x) dx \right|^2 \leq \int_a^b [g(x) - r(x)]^2 dx \int_a^b dx < (b-a) \cdot \varepsilon^2,$$

and thus

$$\left| \int_a^b g(x) dx - \int_a^b r(x) dx \right| = O(\varepsilon). \quad (2.12)$$

So we have

$$J(r,r;u_r) = J(g + (r-g), g + (r-g); u_r)$$
$$= J(g,g;u_r) + J(r-g, r-g; u_r) + 2J(g, r-g; u_r).$$

We can easily bound each term: the first one satisfies

$$J(g,g;u_r) \leq \mu_g$$

due to Eq. (2.10). We can bound the second one as follows:

$$J(r-g, r-g; u_r) = \int_a^b \int_a^b K(t,x)[(g-r)(t)][(g-r)(x)]u_r(t)u_r(x) dxdt$$
$$\leq \left[\int_a^b \int_a^b [K(t,x)u_r(t)u_r(x)]^2 dxdt \right]^{\frac{1}{2}}$$
$$\cdot \left[\int_a^b \int_a^b [(g-r)(t)]^2 [(g-r)(x)]^2 dxdt \right]^{\frac{1}{2}}$$

$$= \left[\int_a^b \int_a^b [K(t,x)u_r(t)u_r(x)]^2 dx dt \right]^{\frac{1}{2}} \cdot \int_a^b [(g-r)(x)]^2 dx$$

$$< \left[\int_a^b \int_a^b [K(t,x)u_r(t)u_r(x)]^2 dx dt \right]^{\frac{1}{2}} \cdot \varepsilon^2$$

$$= C \cdot \varepsilon^2,$$

that is, $J(g-r,g-r;u_r) = O(\varepsilon^2)$.

Finally, the last one is bounded with the aid of Eq. (2.9) and the previous step:

$$J^2(g,g-r;u_r) \leq J(g,g;u_r) \cdot J(g-r,g-r;u_r)$$
$$\leq \mu_g \cdot J(g-r,g-r;u_r)$$
$$= O(\varepsilon^2).$$

Collecting the bounds, we obtain

$$\mu_r \leq \mu_g + O(\varepsilon^2) + O(\varepsilon),$$

and by interchanging the roles of g and r,

$$\mu_g \leq \mu_r + O(\varepsilon^2) + O(\varepsilon).$$

Therefore,

$$|\mu_r - \mu_g| = O(\varepsilon). \tag{2.13}$$

Clearly, the density of the set R in M, together with inequalities (2.12) and (2.13), implies

$$\inf_{g \in M} \left\{ \mu_g^{-\frac{1}{2}} \int_a^b g(x) dx \right\} = \inf_{r \in R} \left\{ \mu_r^{-\frac{1}{2}} \int_a^b r(x) dx \right\}.$$

Now we show that it is enough to consider step functions with at most one discontinuity. To this end, given a nonincreasing function $r \in R$, let $a < x_1 < x_2 < \cdots < x_m = b$ be the discontinuity points of r. We can now write

$$r(x) = \sum_{j=1}^m c_j s_j(x),$$

where

$$s_j = \frac{\chi_{[a,x_j]}(x)}{x_j - a}.$$

and some coefficients $c_j \geq 0$.

Hence

$$\int_a^b r(x) dx = \int_a^b \sum_{j=1}^m c_j s_j(x) dx = \sum_{j=1}^m c_j \int_a^b s_j(x) dx = \sum_{j=1}^m c_j. \tag{2.14}$$

Let us call

$$\mu_s = \max_{1 \le j \le m} \{\mu_{s_j}\},$$

the maximum of the eigenvalues μ_{s_j} of problem (2.8) with weight $g = s_j$.

Let us investigate the relationship between μ_r and μ_s. We have

$$\mu_r = J(r, r; u_r)$$

$$= \sum_{i=1}^{m} \sum_{j=1}^{m} c_i c_j J(s_i, s_j; u_r)$$

$$\le \sum_{i=1}^{m} \sum_{j=1}^{m} c_i c_j [J(s_i, s_i; u_r) J(s_j, s_j; u_r)]^{\frac{1}{2}}$$

$$= \left(\sum_{j=1}^{m} c_j [J(s_j, s_j; u_r)]^{\frac{1}{2}} \right)^2$$

$$\le \left(\sum_{j=1}^{m} c_j \mu_{s_j}^{\frac{1}{2}} \right)^2$$

$$\le \left(\mu_s^{\frac{1}{2}} \int_a^b r(x) dx \right)^2,$$

where we have used Eq. (2.14) in the last step.

Thus,

$$\mu_r^{\frac{1}{2}} \le \mu_s^{\frac{1}{2}} \int_a^b r(x) dx.$$

Finally, since $\int_a^b s(x) dx = 1$, we have

$$\mu_s^{-\frac{1}{2}} \int_a^b s(x) dx \le \mu_r^{-\frac{1}{2}} \int_a^b r(x) dx,$$

which implies that

$$\inf_{s \in S} \left\{ \mu_s^{-\frac{1}{2}} \int_a^b s(x) dx \right\} \le \inf_{r \in R} \left\{ \mu_r^{-\frac{1}{2}} \int_a^b r(x) dx \right\},$$

and the lemma is proved, since $S \subset R$ implies the other inequality.   □

Remark 3.4. This lemma of Nehari proved in [88] cannot be used as stated for p-Laplacian problems, although it can be easily modified, due to its abstract nature. Observe that it is not necessary to know the operator explicitly.

Now we are ready to prove Theorem C, following the original proof in [88]:

Proof. Let λ_w be the first eigenvalue of

$$-u'' = \lambda w(x) u, \qquad x \in (a, b),$$

with zero Dirichlet boundary conditions.

Using the corresponding Green's function, we can consider the equivalent inverse problem, and let $\lambda_w^{-1} = \mu_w$ be the first eigenvalue of problem (2.6).

Now due to Lemma 3.2,

$$\inf_{w \in M} \left\{ \mu_w^{-\frac{1}{2}} \int_a^b w^{\frac{1}{2}}(x) dx \right\} = \inf_{s \in S} \left\{ \lambda_s^{\frac{1}{2}} \int_a^b s^{\frac{1}{2}}(x) dx \right\}$$

can be approached in S, the step functions with at most one discontinuity.

So it is enough to work with w given by

$$w(x) = \begin{cases} \alpha^2 & \text{if } x \in [a, t_1], \\ \beta^2 & \text{if } x \in (t_1, b], \end{cases}$$

for some $t_1 \in (a, b)$, and $\alpha, \beta \geq 0$. Let us set $\gamma = \lambda_w^{\frac{1}{2}}$, and we wish to minimize the function

$$\lambda_w^{\frac{1}{2}} \int_a^b w^{\frac{1}{2}}(x) dx = \gamma \int_a^{t_1} \alpha \, dx + \gamma \int_{t_1}^b \beta \, dx = \gamma \alpha (t_1 - a) + \gamma \beta (b - t_1).$$

Moreover, let us note that the first eigenfunction u_w associated to λ_w satisfies the following problems:

Problem 1: For $t \in [a, t_1]$,

$$v'' + \gamma^2 \alpha^2 v = 0,$$
$$v(a) = 0.$$

Problem 2: For $t \in (t_1, b]$,

$$v'' + \gamma^2 \beta^2 v = 0,$$
$$v(b) = 0.$$

Both problems can be solved explicitly, and the solutions are

$$u(x)|_{(a, t_1)} = \sin[\gamma \alpha (x - a)],$$
$$u(x)|_{(t_1, b)} = \sin[\gamma \beta (b - x)],$$

for $\alpha, \beta > 0$. We paste them smoothly at $t = t_1$, imposing the conditions

$$\sin[\gamma \alpha (t_1 - a)] = c \sin[\gamma \beta (b - t_1)],$$
$$\gamma \alpha \cos[\gamma \alpha (t_1 - a)] = -\gamma c \beta \cos[\gamma \beta (b - t_1)],$$

which implies

$$\beta \tan[\gamma \alpha (t_1 - a)] + \alpha \tan[\gamma \beta (b - t_1)] = 0. \tag{2.15}$$

Using the positivity of the first eigenfunction, we have $\gamma \alpha (t_1 - a) \in (0, \pi)$, and also $\gamma \beta (b - t_1) \in (0, \pi)$. Moreover, from $\alpha + \beta > 0$, Eq. (2.15) implies that

$$\gamma \alpha (t_1 - a) \in (0, \pi/2) \quad \text{and} \quad \gamma \beta (b - t_1) \in (\pi/2, \pi),$$

or

$$\gamma\alpha(t_1 - a) \in (\pi/2, \pi) \quad \text{and} \quad \gamma\beta(b - t_1) \in (0, \pi/2).$$

In both cases, since one of the terms is greater than $\pi/2$, we get

$$\lambda_w^{\frac{1}{2}} \int_a^b w^{\frac{1}{2}}(x)\mathrm{d}x = \gamma\alpha(t_1 - a) + \gamma\beta(b - t_1) > \frac{\pi}{2}.$$

When $\alpha = 0$ or $\beta = 0$, we get $u(x)|_{(a,t_1)} = x - a$ or $u(x)|_{(t_1,b)} = b - x$. As before, we obtain

$$\gamma\beta(t_1 - a) + \tan[\gamma\beta(b - t_1)] = 0,$$
$$\gamma\alpha(b - t_1) + \tan[\gamma\alpha(t_1 - a)] = 0,$$

depending on which parameter was zero. However, in both cases, the first term is positive, so the argument must be greater than $\pi/2$.

Finally, if $t_1 = a$ or $t_1 = b$, we have $s(x) \equiv \beta^2$ in $[a, b]$ and

$$y(t) = \sin[\gamma\beta(t - a)].$$

Since $y(b) = 0$, then $\gamma\beta(b - a) = k\pi$ for some $k \in N$, and the first eigenvalue is

$$\lambda_s = \left(\frac{\pi}{\beta(b - a)}\right)^2.$$

Hence,

$$\lambda_s^{\frac{1}{2}} \int_a^b \sqrt{s(x)}\mathrm{d}x = \frac{\pi}{\beta(b - a)} \int_a^b \beta\mathrm{d}x = \pi > \frac{\pi}{2}.$$

The proof is finished.   □

3.2.2 Nehari's Proof for Linear Higher-Order Differential Equations

The extension to higher-order ordinary differential equations was sketched in [88] only for fourth-order problems. The following theorem gives a complete generalization.

Theorem 3.3. *Let λ_w denote the first eigenvalue of the problem*

$$(-1)^m u(2m) - \lambda w(x)u = 0,$$
$$u(a) = u'(a) = \cdots = u^{m-1}(a) = 0, \tag{2.16}$$
$$u(b) = u'(b) = \cdots = u^{m-1}(1) = 0.$$

Then

$$\inf_{g \in M} \left\{ \lambda_g^{\frac{1}{2m}} \int_a^b \sqrt[2m]{g(x)} dx \right\} = \inf_{s \in S} \left\{ \lambda_s^{\frac{1}{2m}} \int_a^b \sqrt[2m]{s(x)} dx \right\}, \qquad (2.17)$$

where

$$M = \{g : g \in L^2([a,b]) \text{ nonnegative and monotonic}\},$$

$$R = \{r : r \in L^2([a,b]) \text{ monotonic step function}\},$$

$$S = \{s : s \in R \text{ with at most a single discontinuity point in } [a,b]\}.$$

Proof. Since the case $m = 1$ was proved before, we assume that $m > 1$.

Let $G(x,y)$ be the Green's function corresponding to $L(y) = (-1)^m y^{(2m)}$ with Dirichlet boundary conditions. We can write problem (2.16) as an integral equation

$$v_1(x) = \lambda_w \int_a^b G(x,y) \sqrt{w(x)} \sqrt{w(y)} v_1(y) dy \qquad (2.18)$$

by introducing $v_1(x) = u_w(x) \sqrt{w(x)}$, u_w being the first eigenfunction of problem (2.16).

Let us fix $\varepsilon > 0$, and we have

$$\sqrt{w(x)} \sqrt{w(y)} = \left[\sqrt[2m]{w(x)} \sqrt[2m]{w(y)} \right]^{m-1} \sqrt[2m]{w(x)} \sqrt[2m]{w(y)},$$

so we can introduce the auxiliary kernel

$$K_1(x,y) = G(x,y) \left[\sqrt[2m]{w(x)} \sqrt[2m]{w(y)} \right]^{m-1},$$

obtaining from Eq. (2.18) the following problem:

$$\mu_1 v_1(x) = \int_a^b K_1(x,y) \sqrt[2m]{w(x)} \sqrt[2m]{w(y)} v_1(y) dy, \qquad (2.19)$$

where $\mu_1 = \lambda_w^{-1}$. Applying to this problem Lemma 3.2 with $g = \sqrt[2m]{w}$, we get

$$\inf_{w \in M} \left\{ \mu_1^{-\frac{1}{2}} \int_a^b \sqrt[2m]{w(x)} dx \right\} = \inf_{s \in S} \left\{ \mu_s^{-\frac{1}{2}} \int_a^b \sqrt[2m]{s(x)} dx \right\}.$$

Let us take s_1 such that

$$\left| \mu_{s_1}^{-\frac{1}{2}} \int_a^b \sqrt[2m]{s_1(x)} dx - \inf_{s \in S, K_1} \left\{ \mu_s^{-\frac{1}{2}} \int_a^b \sqrt[2m]{s(x)} dx \right\} \right| < \frac{\varepsilon}{m}.$$

We have a new eigenvalue problem now, namely

$$\mu_2 v_2(x) = \int_a^b G(x,y) \left[\sqrt[2m]{w(x)} \sqrt[2m]{w(y)} \right]^{m-1} \sqrt[2m]{s_1(x)} \sqrt[2m]{s_1(y)}) v_2(y) dy$$

$$= \int_a^b \left\{ G(x,y) \left[\sqrt[2m]{w(x)} \sqrt[2m]{w(y)} \right]^{m-2} \sqrt[2m]{s_1(x)} \sqrt[2m]{s_1(y)} \right\}$$

$$\times \sqrt[2m]{w(x)} \sqrt[2m]{w(y)} v_!(y) dy$$

$$= \int_a^b K_2(x,y) \sqrt[2m]{w(x)} \sqrt[2m]{w(y)} v_1(y) dy,$$

where $K_2(x,y) = G(x,y) \left[\sqrt[2m]{w(x)} \sqrt[2m]{w(y)} \right]^{m-2} \sqrt[2m]{s_1(x)} \sqrt[2m]{s_1(y)}$.

This problem is similar to the previous one; we have only changed the kernel, and the same argument shows that there exists some function $s_2 \in S$ such that

$$\left| \mu_{s_2}^{-\frac{1}{2}} \int_a^b \sqrt[2m]{s_2(x)} dx - \inf_{s \in S, K_2} \left\{ \mu_s^{-\frac{1}{2}} \int_a^b \sqrt[2m]{s(x)} dx \right\} \right| < \frac{\varepsilon}{m}.$$

In this way, we define a sequence of problems for $1 \leq i \leq m$,

$$\mu_i v_i v(x) = \int_a^b K_i(x,y) \sqrt[2m]{w(x)} \sqrt[2m]{w(y)} v_i(y) dy, \qquad (2.20)$$

where

$$K_i(x,y) = \left[\sqrt[2m]{w(x)} \sqrt[2m]{w(y)} \right]^{m-i} \cdot \prod_{k=1}^{i-1} \sqrt[2m]{s_k(x)} \sqrt[2m]{s_k(y)},$$

and each function s_k, $1 \leq k \leq i-1$, is a simple monotonic function with a single discontinuity. From Lemma 3.2, we obtain a new function s_i such that

$$\left| \mu_{s_i}^{-\frac{1}{2}} \int_a^b \sqrt[2m]{s_i(x)} dx - \inf_{s \in S, K_i} \left\{ \mu_s^{-\frac{1}{2}} \int_a^b \sqrt[2m]{s(x)} dx \right\} \right| < \frac{\varepsilon}{m}. \qquad (2.21)$$

Finally, in the last step we have

$$\mu_m v_m(x) = \int_a^b K_m(x,t) \sqrt[2m]{s_m(x)} \sqrt[2m]{s_m(y)} v_m(y) dy, \qquad (2.22)$$

where $\sqrt[2m]{s_m}$ is as before, and

$$K_m(x,y) = G(x,y) \prod_{k=1}^{m-1} \sqrt[2m]{s_k(x)} \sqrt[2m]{s_k(y)}.$$

Hence

$$\left| \mu_{s_m}^{-\frac{1}{2}} \int_a^b \sqrt[2m]{s_m(x)} dx - \inf_{w \in M} \left(\mu_w^{-\frac{1}{2}} \int_a^b \sqrt[2m]{w(x)} dx \right) \right| < \varepsilon,$$

by rearranging the terms in

$$\left| \mu_{s_m}^{-\frac{1}{2}} \int_a^b \sqrt[2m]{s_m(x)} dx - \left(\sum_{i=1}^{m-1} \mu_{s_i}^{-\frac{1}{2}} \int_a^b \sqrt[2m]{s_i(x)} dx \right) \right.$$

$$\left. + \left(\sum_{i=1}^{m-1} \mu_{s_i}^{-\frac{1}{2}} \int_a^b \sqrt[2m]{s_i(x)} dx \right) - \inf_{w \in M} \left(\mu_w^{-\frac{1}{2}} \int_a^b \sqrt[2m]{w(x)} dx \right) \right|$$

and using the inequalities (2.21).

However, μ_{s_m} is the first eigenvalue of problem

$$\mu v(x) = \int_a^b G(x,y) \prod_{i=1}^m \sqrt[2m]{s_i(x)} \sqrt[2m]{s_i(y)} v(y) dy,$$

and the weight is a simple monotonic function, but we cannot guarantee that it has a unique discontinuity.

So let v_m be the normalized eigenfunction corresponding to μ_m. Then

$$\mu_m = \int_a^b \int_a^b G(x,y) \prod_{i=1}^m \sqrt[2m]{s_i(x)} \sqrt[2m]{s_i(y)} v_m(x) v_m(y) dy dx. \qquad (2.23)$$

Since G and u_m are nonnegative, we can write

$$G(x,y) v_m(x) v_m(y) = [G(x,y) v_m(x) v_m(t)]^{\frac{1}{m}} \cdots [G(x,y) v_m(x) v_m(t)]^{\frac{1}{m}},$$

and we get

$$\mu_m = \int_a^b \int_a^b \prod_{i=1}^m \left[G(x,y) v_m(x) v_m(y) \sqrt{s_i(x)} \sqrt{s_i(y)} \right]^{\frac{1}{m}} dx dy.$$

From Hölder's inequality we obtain

$$\mu_m^m \leq \prod_{i=1}^m \int_a^b \int_a^b G(x,y) v_m(x) v_m(y) \sqrt{s_i(x)} \sqrt{s_i(y)} dx dy. \qquad (2.24)$$

We can consider one of the problems

$$\mu v(x) = \int_a^b G(x,y) \sqrt{s_k(x)} \sqrt{s_k(y)} v(y) dy,$$

which is equivalent to the differential problem

$$(-1)^m u^{(2m)} - \lambda s_k(x) u = 0,$$
$$u(a) = u'(a) = \cdots = u^{m-1}(a) = 0, \qquad (2.25)$$
$$u(b) = u'(b) = \cdots = u^{m-1}(b) = 0.$$

The first eigenvalue is $\lambda_{s_k} = \mu_k^{-1}$, and the eigenfunction u_{s_k} satisfies $v_k = \sqrt{s_k} u_{s_k}$.
Since

$$\mu_k = J(\sqrt{s_k}, \sqrt{s_k}; v_k) \geq J(\sqrt{s_k}, \sqrt{s_k}; u_{s_m}),$$

we obtain in Eq. (2.24)

$$\mu_{s_m}^m \leq \prod_{k=1}^{m} \mu_k \leq \max\{\mu_1, \mu_2, \ldots, \mu_m\}^m.$$

Therefore,

$$\mu_{s_m} \leq \max\{\mu_1, \mu_2, \ldots, \mu_m\} = \mu_s$$

for some simple function s with at most one discontinuity, and the theorem is
proved.   □

3.3 The Inequality for p-Laplacian Problems

We start with the eigenvalue problem

$$-(|u'|^{p-2}u')' = \lambda w(x)|u|^{p-2}u, \qquad u(a) = u(b) = 0, \tag{3.26}$$

and let us recall that the first eigenvalue λ_w can be obtained as

$$\lambda_w^{-1} = \sup_{\{u \in W_0^{1,p}(a,b): \int_a^b |u'|^p dx = 1\}} \int_a^b w(x)|u|^p dx.$$

Theorem 3.4. *Let $w \in L^1(a,b)$ be a positive monotonic weight, and λ_w the first
eigenvalue of problem (3.26). Then*

$$\frac{\pi_p}{2} \leq \lambda_w^{\frac{1}{p}} \int_a^b w^{\frac{1}{p}}(x) dx.$$

We need an alternative formulation of Lemma 3.2. Basically, this lemma says that
the eigenvalues are close when the weights are close, and also that you can change a
simple monotonic function by a step function with at most one discontinuity without
increasing the eigenvalue.

So this is our first objective: to prove some approximation result for the eigenvalues whenever some norm of the difference of the weights is small.

Lemma 3.3. *Given the eigenvalue problem (3.26), where $w \in L^1(a,b)$ is a positive
monotonic function, and $\varepsilon > 0$, there exists a simple monotonic weight $r(x)$ such
that*

$$(i) \quad |\lambda_w - \lambda_r| = O(\varepsilon),$$

$$(ii) \quad \left| \int_a^b w^{\frac{1}{p}}(x) dx - \int_a^b r^{\frac{1}{p}}(x) dx \right| = O(\varepsilon),$$

where λ_r is the first eigenvalue of

$$-(|u'|^{p-2}u')' = \lambda r(x)|u|^{p-2}u, \qquad u(a) = u(b) = 0.$$

Proof. Let $r \in L^1(a,b)$ such that $\|w - r\|_{L^1} < \varepsilon$. Let u_w, u_r be the normalized (such that $\|u'_w\|_p = \|u'_r\|_p = 1$) associated eigenfunctions to the first eigenvalues λ_w, λ_r. Let us recall first that the compactness of the embedding $W_0^{1,p}(a,b) \hookrightarrow L^\infty(a,b)$ implies that both eigenfunctions are uniformly bounded in $L^\infty(a,b)$ by a fixed constant C. Hence,

$$\begin{aligned}
\lambda_w^{-1} &= \int_a^b w(x)u_w^p(x)\mathrm{d}x \\
&= \int_a^b r(x)u_w^p(x)\mathrm{d}x + \int_a^b (w(x) - r(x))u_w^p(x)\mathrm{d}x \\
&\leq \lambda_r^{-1} + \|u_w\|_{L^\infty}^p \int_a^b |w(x) - r(x)|\mathrm{d}x \\
&\leq \lambda_r^{-1} + C^p\varepsilon.
\end{aligned}$$

The same argument, starting from λ_r, gives

$$\lambda_r^{-1} \leq \lambda_w^{-1} + C^p\varepsilon,$$

and (i) is proved.

Minkowski's inequality implies that

$$\int_a^b |w^{\frac{1}{p}}(x) - r^{\frac{1}{p}}(x)|\mathrm{d}x \leq \int_a^b |w(x) - r(x)|^{\frac{1}{p}}\mathrm{d}x,$$

which can be bounded by $(b-a)^{\frac{p-1}{p}}\|w - r\|_1^{\frac{1}{p}}$ by Hölder's inequality. So (ii) is proved. $\qquad\square$

Remark 3.5. As usual, a Sturmian argument enables us to work with arbitrary monotonic functions, changing w to w^+.

Remark 3.6. It follows that given a positive monotonic function $w \in L^1(a,b)$ and $\varepsilon > 0$, we can take a simple function r such that

$$\lambda_w^{\frac{1}{p}} \int_a^b w^{\frac{1}{p}}(x)\mathrm{d}x = \lambda_r^{\frac{1}{p}} \int_a^b r^{\frac{1}{p}}(x)\mathrm{d}x + O(\varepsilon).$$

Our next task is to show that we can replace r by a step function with at most a single discontinuity point.

Lemma 3.4. *Given a positive monotonic simple function r, there exists a step function s with at most a single discontinuity point such that*

$$\int_a^b r^{\frac{1}{p}}(x)\mathrm{d}x = \int_a^b s^{\frac{1}{p}}(x)\mathrm{d}x,$$

and $\lambda_s \leq \lambda_r$.

Proof. We write $r^{\frac{1}{p}} = \sum_{j=1}^{n} c_j s_j^{\frac{1}{p}}$ as before, where each function s_j has at most one discontinuity, rescaled such that

$$\int_a^b r^{\frac{1}{p}}(x)\mathrm{d}x = \int_a^b s_j^{\frac{1}{p}}(x)\mathrm{d}x.$$

In particular, we have now

$$\sum_{j=1}^{n} c_j = 1,$$

which enables us to apply Jensen's inequality, with u_r an eigenfunction corresponding to λ_r satisfying $\int_a^b |u'|^p \mathrm{d}x = 1$:

$$\begin{aligned}
\lambda_r^{-1} &= \int_a^b r(x)u_r^p(x)\mathrm{d}x \\
&= \int_a^b \left(\sum_{j=1}^{n} c_j s_j^{\frac{1}{p}}(x) \right)^p u_r^p(x)\mathrm{d}x \\
&\leq \int_a^b \sum_{j=1}^{n} c_j s_j(x)u_r^p(x)\mathrm{d}x \\
&\leq \sum_{j=1}^{n} c_j \lambda_{s_j}^{-1} \\
&\leq \max_{1\leq j\leq n} \{\lambda_{s_j}^{-1}\} \\
&= \lambda_s^{-1},
\end{aligned}$$

where s is the function that has the greater eigenvalue among the functions $\{s_j\}_{j=1}^{n}$. Clearly,

$$\lambda_s^{\frac{1}{p}} \int_a^b s^{\frac{1}{p}}(x) \leq \lambda_r^{\frac{1}{p}} \int_a^b r^{\frac{1}{p}}(x)\mathrm{d}x,$$

and the lemma is proved.   □

Remark 3.7. Observe that this proof is almost identical to the previous one of Nehari. The only difference is that the normalization of the eigenfunction was made with the term corresponding to the operator, that is, $\int_a^b |u'|^p \mathrm{d}x = 1$, instead of the normalization in L^2 with $\int_a^b u^2 \mathrm{d}x = 1$.

Due to this fact, Lemma 3.4 does not depend on the operator, and it can be applied to a wide class of eigenvalue problems.

Remark 3.8. Indeed, a variant of this lemma was proved first by Nehari himself in [89]. He applied it to linear higher-order problems. Basically, we can abstract the following key ideas:

- Convexity of some class of functions R.
- Density of finite combinations of extremals S.

- Convexity of the functional I:

$$I(s) \leq \sum c_j I(s_j),$$

for $s = \sum c_j s_j$ with $\sum c_j = 1$.
- Lower semicontinuity of the functional I, that is,

$$I(r) \leq \liminf_{s \to r} I(s).$$

And the method can be applied to other extremal problems satisfying this structure.

In order to prove Theorem D for p-Laplacian operators, we have only to find a lower bound for $\lambda_s^{\frac{1}{p}} \int_a^b s^{\frac{1}{p}}(x)$ not depending on s. Of course, we can do it by pasting the solutions of constant-coefficient problems, but this is not easy for higher-order operators. This is our next lemma:

Lemma 3.5. *Let s a nonnegative step function with at most one discontinuity in $[a,b]$. Then*

$$\frac{\pi_p}{2} < \lambda_s^{\frac{1}{p}} \int_a^b s^{\frac{1}{p}}(x) dx.$$

Proof. Again, it is enough to work with s given by

$$s(x) = \begin{cases} \alpha^p & \text{if } x \in [a,t_1], \\ \beta^p & \text{if } x \in (t_1,b], \end{cases}$$

for some $t_1 \in (a,b)$, and $\alpha, \beta \geq 0$.

Using the weak formulation of the problem, with u_s the first eigenfunction as a test function, we now have

$$\int_a^b u_s'^p(x) dx = \lambda_s \int_a^b s(x) u_s^p(x) dx,$$

and by splitting the integrals, we obtain

$$\left[\int_a^{t_1} u_s'^p(x) dx - \lambda_s \alpha^p \int_a^{t_1} u_s^p(x) dx \right] + \left[\int_{t_1}^b u_s'^p(x) dx - \lambda_s \beta^p \int_{t_1}^b u_s^p(x) dx \right] = 0.$$

It is clear that one of the terms must be nonpositive, and let us assume that the first one satisfies

$$\int_a^{t_1} u_s'^p(x) dx - \lambda_s \alpha^p \int_a^{t_1} u_s^p(x) dx \leq 0.$$

Therefore,

$$\lambda_s^{-1} \alpha^{-p} \leq \frac{\int_a^{t_1} u_s^p(x) dx}{\int_a^{t_1} u_s'^p(x) dx}.$$

Let us note that u_s is in W, the Sobolev space

$$W = \{ u \in W^{1,p}([a,t_1]) : u(a) = 0 \},$$

so it is an admissible function in the variational characterization of the first eigenvalue of the following problem with mixed boundary conditions:

$$-(|u'|^{p-2}u')' = \lambda |u|^{p-2}u \quad u(a) = u'(b) = 0,$$

that is,

$$\lambda = \inf_{u \in W} \frac{\int_a^{t_1} u^p(x)dx}{\int_a^{t_1} u'^p(x)dx},$$

which can be computed explicitly,

$$\lambda = \left(\frac{\pi_p}{2(t_1 - a)} \right)^p;$$

see Remark A.2 in Appendix A, and then

$$\lambda_s^{-1} \alpha^{-p} \leq \left(\frac{2(t_1 - a)}{\pi_p} \right)^p.$$

Hence

$$\frac{\pi_p}{2} \leq \lambda_s^{\frac{1}{p}} \int_a^{t_1} \alpha dx \leq \lambda_s^{\frac{1}{p}} \int_a^b s^{\frac{1}{p}}(x)dx,$$

and the proof is finished.   □

We are ready to prove the main theorem of this chapter.

Proof (Proof of Theorem 3.4). The proof is almost finished, we have only to collect the previous results: given $w \in L^1(a,b)$ a positive monotonic function, we have

$$\frac{\pi_p}{2} \leq \lambda_w^{\frac{1}{p}} \int_a^b w^{\frac{1}{p}}(x)dx + O(\varepsilon),$$

and Theorem 3.4 is proved.   □

3.3.1 An Extension for Different Powers

Let us observe that in Lemma 3.3, we have

$$\left| \int_a^b w^{\frac{1}{q}}(x)dx - \int_a^b r^{\frac{1}{q}}(x)dx \right| = O(\varepsilon)$$

for every fixed $q \in (1, \infty)$, not only for $q = p$.

Also, Lemma 3.4 holds for $q \in (1, \infty)$: there exists some step function s with at most one discontinuity such that

$$\int_a^b r^{\frac{1}{q}}(x)dx = \int_a^b s^{\frac{1}{q}}(x)dx$$

and $\lambda_s \leq \lambda_r$, since we have

$$r(x) = \left(\sum_{j=1}^{n} c_j s_j^{\frac{1}{q}}(x) \right)^q \leq \sum_{j=1}^{n} c_j s_j(x),$$

due to Jensen's inequality, and the rest of the proof follows as before. Hence, with those easy changes in the proof, we get

$$\lambda_s^{\frac{1}{q}} \int_a^b s^{\frac{1}{q}}(x) \leq \lambda_r^{\frac{1}{q}} \int_a^b r^{\frac{1}{q}}(x)dx,$$

where p and q are not necessarily related.

We now wonder whether we can bound $\lambda_s^{\frac{1}{q}} \int_a^b s^{\frac{1}{q}}(x)$ from below with a constant independent of the length of the interval as in Lemma 3.5.

Here, we must consider separately nonincreasing and nondecreasing weights. Let us fix $q > 1$ and r a nonincreasing simple function. We can write r as the sum of step functions of the form

$$s(x) = \begin{cases} \alpha^q & \text{if } x \in [a, t_1], \\ 0 & \text{if } x \in (t_1, b], \end{cases}$$

for some $t_1 \in (a, b)$, and $\alpha > 0$.

Since the weight s in (t_1, b) is zero, the solution u_w must be a linear function in this interval, and since the derivative of the eigenfunction has a zero $c \in (a, b)$, we must have $c \in (a, t_1)$. Observe that we can also apply an extension for p-Laplacian problems of Theorem 2.3 from Chap. 2: if $c \in (t_1, b)$, we must have

$$1 \leq (b - c)^{p-1} \sup_{c \leq x \leq b} \int_c^x w(s)ds = 0,$$

a contradiction.

Now the same argument used in the proof of Lemma 3.5 (or the monotonicity of the eigenvalues with respect to the domain) implies

$$\lambda_s^{-1} \alpha^{-q} \leq \left(\frac{2(t_1 - a)}{\pi_p} \right)^p,$$

that is,

$$\lambda_s^{-\frac{1}{q}} \alpha^{-1} \leq \left(\frac{2(t_1 - a)}{\pi_p} \right)^{\frac{p}{q}},$$

and we get

$$\left(\frac{\pi_p}{2}\right)^{\frac{p}{q}} \leq \lambda_s^{\frac{1}{q}} (t_1 - a)^{\frac{p}{q}} \alpha$$

$$= \frac{p}{q} \lambda_s^{\frac{1}{q}} \int_a^{t_1} (x-a)^{\frac{p}{q}-1} \alpha dx$$

$$\leq \frac{p}{q} \lambda_s^{\frac{1}{q}} \int_a^b (x-a)^{\frac{p}{q}-1} s^{\frac{1}{q}}(x) dx.$$

We have proved the following theorem:

Theorem 3.5. *Let* $w \in L^1(a,b)$ *be a monotonic weight, and* λ_w *the first eigenvalue of problem (3.26). Then if* w *is nonincreasing, we have*

$$\frac{q}{p} \left(\frac{\pi_p}{2}\right)^{\frac{p}{q}} \leq \lambda_w^{\frac{1}{q}} \int_a^b (x-a)^{\frac{p-q}{q}} w^{\frac{1}{q}}(x) dx,$$

and if w *is nondecreasing, then*

$$\frac{q}{p} \left(\frac{\pi_p}{2}\right)^{\frac{p}{q}} \leq \lambda_w^{\frac{1}{q}} \int_a^b (b-x)^{\frac{p-q}{q}} w^{\frac{1}{q}}(x) dx.$$

3.4 Optimality of the Bound

The optimality of the constant $\pi/2$ was mentioned by Cohn and Calogero without proof, and Nehari obtained it as a consequence of the bound with a weight $w(x) \equiv 1$ for $x \in (0,t_1)$ and $w(x) \equiv 0$ for $x \in (t_1, \pi)$. The continuity and differentiability of the solution imply

$$\gamma(\pi - t_1) + \tan[\gamma t_1] = 0,$$

and the solution can be computed explicitly,

$$t_1 = \left(\frac{\pi}{2} + \varepsilon\right) \left(\frac{1}{2} + \frac{\varepsilon + \cot(\varepsilon)}{\pi}\right)^{-1},$$

$$\gamma = \frac{1}{2} + \frac{\varepsilon + \cot(\varepsilon)}{\pi}.$$

Hence

$$\lambda_w^{\frac{1}{2}} \int_0^{\pi} w^{\frac{1}{2}}(x) dx = \gamma t_1 = \frac{\pi}{2} + \varepsilon.$$

A similar argument holds for $p \neq 2$.

3.5 Higher Eigenvalues and Some Nonmonotonic Weights

We present here an inequality for higher eigenvalues and an extension for nonmonotonic weights with finitely many changes of monotonicity.

3.5.1 Higher Eigenvalues

The following theorem was obtained by Nehari in [88] for the linear problem when
$p = 2$, and both Calogero and Cohn have used the same idea in their works.

Theorem 3.6. *Let λ_n be the nth eigenvalue of*

$$- (|u'|^{p-2}u')' = \lambda w(x)|u|^{p-2}u, \qquad u(a) = u(b) = 0, \qquad (5.27)$$

where w is a positive monotonic function. Then

$$\frac{n^p \pi_p^p}{2^p} < \lambda_n \left(\int_a^b w^{\frac{1}{p}}(x)dx \right)^p.$$

Proof. We use the fact that the eigenfunction u_n corresponding to λ_n has $n+1$ simple zeros,

$$a = x_0 < x_1 < \cdots < x_n = b.$$

We can apply now Theorem C in each interval (x_{i-1}, x_i), obtaining

$$n \cdot \frac{\pi_p}{2} < \sum_{i=1}^{n} \lambda_n^{\frac{1}{p}} \int_{x_{i-1}}^{x_i} w^{\frac{1}{p}}(x)dx = \lambda_n^{\frac{1}{p}} \int_a^b w^{\frac{1}{p}}(x)dx,$$

and the result follows by raising both members to the power p. $\qquad\qquad \square$

Remark 3.9. The argument in the works of Calogero and Cohn [17, 18, 28, 29] to
derive an upper bound for the number of negative eigenvalues of one-dimensional
Schrödinger operators is very similar.

Let us consider the problem

$$\psi'' + w(x)\psi = E\psi, \qquad x \geq 0,$$

where $\psi \in H_0^1(0, \infty)$ is the wave function, w is a positive and decreasing potential,
and E is the eigenvalue parameter. Let us assume that $E_n < 0$ is an eigenvalue,
and ψ_n is the associated eigenfunction, which has $n + 1$ zeros. Then the Sturmian
comparison theorem implies that every solution $u \in H_0^1(0, \infty)$ of

$$u'' + w(x)u = 0, \qquad x \geq 0,$$

has at least $n+1$ zeros $\{x_i\}_{i=0}^n \subset [0,\infty)$. Now applying the inequality in each interval as in the previous proof and extending the integral to the half-line gives

$$n < \frac{2}{\pi} \int_0^\infty w^{\frac{1}{2}}(x)\mathrm{d}x.$$

This is the Calogero–Cohn upper limit on the number of bound states.

3.5.2 Nonmonotonic Weights

We consider the case of weights with finitely many changes of monotonicity. In this section we assume that w is a positive weight that is monotonic in k intervals. Although it is not possible to derive a lower bound for all the eigenvalues, we can bound the higher eigenvalues. The following theorem extends the one proved by Nehari [88] and Cohn [28] to p-Laplacian operators.

Theorem 3.7. *Let λ_n be the nth eigenvalue of*

$$-(|u'|^{p-2}u')' = \lambda w(x)|u|^{p-2}u, \qquad u(a) = u(b) = 0,$$

where w is a positive function, $w^{\frac{1}{p}} \in L^1(a,b)$, which is monotonic in $k+1$ intervals (y_{i-1}, y_i), for $1 \leq i \leq k+1$, and $y_0 = a$, $y_{k+1} = b$. Then for $n > k$, we have

$$\frac{(n-k)^p \pi_p^p}{2^p} < \lambda_n \left(\int_a^b w^{\frac{1}{p}}(x)\mathrm{d}x \right)^p.$$

Proof. Again, let us recall that the eigenfunction u_n corresponding to λ_n has $n+1$ simple zeros,

$$a = x_0 < x_1 < \cdots < x_n = b.$$

Now we can apply Theorem C in a given interval (x_{i-1}, x_i), provided that $y_j \notin (x_{i-1}, x_i)$, so we can apply it in at least $n-k$ intervals. Of course, we can add in the right-hand side the other intervals, obtaining

$$(n-k) \cdot \frac{\pi_p}{2} < \sum_{i=1}^n \lambda_n^{\frac{1}{p}} \int_{x_{i-1}}^{x_i} w^{\frac{1}{p}}(x)\mathrm{d}x = \lambda_n^{\frac{1}{p}} \int_a^b w^{\frac{1}{p}}(x)\mathrm{d}x,$$

and the result follows as before by raising both members to the power p. $\qquad\square$

Chapter 4
Bargmann-Type Bounds

Abstract In this chapter we study Bargmann-type bounds for the number of negative eigenvalues of Schrödinger equations, which is a well-known hypothesis for the nonoscillatory behavior of solutions of second-order problems on the half-line. Such bounds mean that there are no solutions with two zeros if $\|xw(x)\|_{L^1}$ is smaller than a certain constant. We present some lower bounds for the first eigenvalue of linear and quasilinear problems involving integrals of the weights and some power of x, and we apply it to singular eigenvalue problems in unbounded intervals.

4.1 Bargmann-Type Bounds

In this section we prove Theorem D, which states that the inequality

$$\frac{\pi}{4} \leq \int_a^b (x-a)w(x)\mathrm{d}x \tag{1.1}$$

must be satisfied in order for there to be a nontrivial solution of

$$-u'' = w(x)u, \qquad u(a) = u(b) = 0. \tag{1.2}$$

As we will see below, we can extend this inequality to higher-order equations and p-Laplacian problems.

We call this kind of inequality a Bargmann-type bound, since Bargmann considered in [4] the stationary radial wave equation

$$-u'' + \frac{l(l+1)}{r^2}u = w(x)u, \tag{1.3}$$

where $l \geq 0$ is the angular moment, and w satisfies

$$\int_0^\infty x|w(x)|\mathrm{d}x < \infty.$$

J.P. Pinasco, *Lyapunov-type Inequalities: With Applications to Eigenvalue Problems*, SpringerBriefs in Mathematics, DOI 10.1007/978-1-4614-8523-0_4, © Juan Pablo Pinasco 2013

There are no other assumptions on the regularity of w, and all kinds of singularities are allowed. As usual, it suffices to consider a positive function w.

Bargmann's main result is the following, which generalizes Theorem D:

Theorem 4.1. *Let u be a solution of Eq. (1.3) in $(0,\infty)$, with $u(0) = 0$ and vanishing in ∞. Let a, b be two consecutive zeros. Then*

$$2l + 1 < \int_a^b x|w(x)|\,dx.$$

First we give a proof of Theorem 4.1, following the original proof of Bargmann [4]. Then we consider an alternative and much simpler proof, with a worse lower bound for $l = 0$, and we will discuss its extensions to p-Laplacian operators.

As an application, let us recall that the well-known inequality

$$\int_0^\infty xw(x)\,dx < \infty$$

implies that every solution of

$$-u'' = \lambda w(x)u, \qquad x \geq 0,$$

has finitely many zeros for every value of λ, that is, u is strongly nonoscillatory. Similar conditions hold for higher-order problems and quasilinear problems. So Theorem D gives a different proof of those nonoscillation results.

4.1.1 Bargmann's Proof

Let us prove the main theorem of this chapter.

Proof (Proof of Theorem 4.1). Let us observe first that the solution of Eq. (1.3), scaled such that

$$\lim_{x \to 0} x^{l+1} u(x) = 1,$$

can be written as

$$u(x) = x^{l+1} - \int_0^x G(x,y)w(y)u(y)\,dy, \qquad (1.4)$$

where G is the Green's function of the operator

$$-u'' + \frac{l(l+1)}{r^2}u = 0,$$

given explicitly by the expression

$$G(x,y) = \frac{1}{2l+1}\left[x\left(\frac{x}{y}\right)^l - y\left(\frac{y}{x}\right)^l\right].$$

Moreover,

$$\lim_{x \to \infty} x^{-(l+1)} u(x) = \gamma,$$

and u is increasing at infinity when $\gamma \neq 0$.

The proof is divided in four cases:

1. Case $a = 0$, $b < \infty$. From the integral representation at $x = b$, since $u(b) = 0$ and $u(y) \leq y^{l+1}$, we get

$$b^{l+1} = \int_0^b G(b,y)w(y)u(y)dy \leq \int_0^b G(b,y)w(y)y^{l+1}dy.$$

Using the expression for G, we have

$$(2l+1)b^{l+1} \leq \int_0^b \left[b\left(\frac{b}{y}\right)^l - y\left(\frac{y}{b}\right)^l \right] w(y)y^{l+1}dy$$

$$= \int_0^b b^{l+1}w(y)ydy - \int_0^b w(y)\frac{y^{2+2}}{b^l}dy$$

$$\leq \int_0^b b^{l+1}w(y)ydy,$$

and the inequality holds in this case.

2. Case $a = 0$, $b = \infty$. This is similar to the previous one. Recall that we have $u(x) \leq x^{l+1}$, and since u vanishes at infinity, then $x^{-(l+1)}u(x)$ goes to 0 as $x \to \infty$. From $u(x) = x^{l+1} - \int_0^x G(x,y)w(y)u(y)dy$, we obtain, after multiplying the equality by $(2l+1)x^{-(l+1)}$,

$$\frac{(2l+1)}{x^{l+1}}u(x) = (2l+1) - \frac{(2l+1)}{x^{l+1}} \int_0^x G(x,y)w(y)u(y)dy.$$

However,

$$\frac{(2l+1)}{x^{l+1}}G(x,y) = y^{-l}\left[1 - \left(\frac{y}{x}\right)^{(2l+1)}\right] \leq y^{-l},$$

which implies, since $u(y) \leq y^{l+1}$,

$$\frac{(2l+1)}{x^{l+1}}u(x) \geq (2l+1) - \int_0^x y^{-l}w(y)y^{l+1}dy$$

$$\geq (2l+1) - \int_0^x w(y)ydy,$$

and the inequality is obtained when $x \to \infty$ since $(2l+1)x^{-(l+1)}u(x) \to 0$.

3. Case $a > 0$, $b < \infty$. We can assume that $u(a) = 0$ and $u'(a) = 1$, multiplying u by a constant if necessary.

We can verify easily that u can be written as

$$u(x) = \frac{1}{2l+1} \left[x \left(\frac{x}{a} \right)^l - a \left(\frac{a}{x} \right)^l \right] - \int_a^x G(x,y)w(y)u(y)dy$$

in (a,b), and in this interval, we have $0 < u(x) < G(x,a)$. Therefore, since $u(b) = 0$, we obtain

$$G(b,a) = \int_a^b G(b,y)w(y)u(y)dy \le \int_a^b G(b,y)w(y)G(y,a)dy.$$

We need now the following inequality:

$$G(b,y)G(y,a) \le \frac{y}{2l+1} G(b,a), \tag{1.5}$$

which can be proved with a simple (but tedious) computation; see Lemma 4.1 below. We have

$$(2l+1)G(b,a) \le G(b,a) \int_a^b yw(y)dy,$$

and we obtain the desired inequality by canceling out $G(b,a)$ on both sides.

4. Case $a > 0$, $b = \infty$. This is similar to the previous case, taking u normalized such that $u'(a) = 1$. Since $u(x) \le G(x,a)$ and u vanishes at infinity,

$$\lim_{x \to \infty} \frac{u(x)}{G(x,a)} = 0.$$

Now, from

$$u(x) = G(x,a) - \int_a^x G(x,y)w(y)u(y)dy,$$

we get

$$(2l+1)\frac{u(x)}{G(x,a)} - (2l+1) + (2l+1) \int_a^x \frac{G(x,y)}{G(x,a)}w(y)u(y)dy = 0.$$

In order to finish the proof, it is enough to show that

$$(2l+1)\frac{G(x,y)}{G(x,a)}w(y)u(y) \le yw(y)$$

and to take the limit $x \to \infty$ as in the second case. To this end, it is enough to use the bound $u(y) \le G(y,a)$ and Lemma 4.1 below,

$$G(x,y)G(y,a) \le \frac{y}{2l+1}G(x,a).$$

Finally, let us observe that the bounds $u(x) \leq x^{l+1}$ and $u(x) \leq G(x,a)$ are strict except at $x = 0$ and $x = a$; if not, we would have $u \equiv 0$ to avoid the contribution of the integral, a contradiction. So all the inequalities are strict, and the proof is finished. $\square$

Remark 4.1. Let us mention that the proofs of cases 2 and 4 are simpler than the original proofs of Bargmann, who used Lemma 4.1 only in case 3 and introduced new auxiliary functions.

4.1.1.1 Proof of Inequality (1.5)

The following lemma was proved in [4]. We give a slightly different and simpler proof.

Lemma 4.1. *Let $G(x,y)$ be the Green's function of the radial wave equation with angular moment l. Then*

$$G(b,y)G(y,a) \leq \frac{y}{2l+1} G(b,a).$$

Proof. Let us observe that is enough to prove the equivalent inequality

$$\left[b\left(\frac{b}{y}\right)^l - y\left(\frac{y}{b}\right)^l \right] \cdot \left[y\left(\frac{y}{a}\right)^l - a\left(\frac{a}{y}\right)^l \right]$$

$$\leq y \left[b\left(\frac{b}{a}\right)^l - a\left(\frac{a}{b}\right)^l \right] - 2y(ab)^{1/2} \left[1 - \left(\frac{a}{b}\right)^{l+1/2} \right].$$

First, we wish to bound the difference

$$y \left[b\left(\frac{b}{a}\right)^l - a\left(\frac{a}{b}\right)^l \right] - \left[b\left(\frac{b}{y}\right)^l - y\left(\frac{y}{b}\right)^l \right] \cdot \left[y\left(\frac{y}{a}\right)^l - a\left(\frac{a}{y}\right)^l \right],$$

and after expanding, we get

$$yb\left(\frac{b}{a}\right)^l - ya\left(\frac{a}{b}\right)^l - yb\left(\frac{b}{a}\right)^l + ab\left(\frac{ab}{y^2}\right)^l + y^2\left(\frac{y^2}{ab}\right)^l - ya\left(\frac{a}{b}\right)^l.$$

Collecting and canceling out some terms, we obtain

$$ab\left(\frac{ab}{y^2}\right)^l + y^2\left(\frac{y^2}{ab}\right)^l - 2ya\left(\frac{a}{b}\right)^l,$$

and rewriting it, we get an expression that is close to a square,

$$\left(\frac{(\sqrt{ab})^{l+1}}{y^l}\right)^2 + \left(\frac{y^{l+1}}{(\sqrt{ab})^l}\right)^2 - 2ya\left(\frac{a}{b}\right)^l,$$

that is,

$$\left(\frac{(\sqrt{ab})^{l+1}}{y^l} - \frac{y^{l+1}}{(\sqrt{ab})^l}\right)^2 + 2\sqrt{ab}\,y - 2ya\left(\frac{a}{b}\right)^l.$$

Finally, we can write this as

$$\left(\frac{(\sqrt{ab})^{l+1}}{y^l} - \frac{y^{l+1}}{(\sqrt{ab})^l}\right)^2 + 2y\sqrt{ab}\left[1 - \left(\frac{a}{b}\right)^{l+1/2}\right],$$

and on discarding the first term, which is positive, the inequality is proved.   □

4.1.2 A Shorter Proof

The following proof is optimal only for $l = 0$. Nevertheless, it gives a quick bound. Starting from

$$-u'' + +\frac{l(l+1)}{r^2}u = w(x)u,$$

with a, b two consecutive zeros, we can assume that $u \geq 0$, and that $u'(c) = 0$ for some $c \in (a, b)$ where u reaches its maximum.

Hence, taking $t \in (a, c)$, we get by direct integration of the equation that

$$u'(t) + \int_t^c \frac{l(l+1)}{s^2}u(s)ds = \int_t^c w(s)u(s)ds,$$

and integrating again, since $u(a) = 0$,

$$\begin{aligned} u(c) &= \int_a^c u'(t)dt \\ &\leq \int_a^c u'(t)dt + \int_a^c \int_t^c \frac{l(l+1)}{s^2}u(s)dsdt \\ &= \int_a^c \int_t^c w(s)u(s)dsdt \\ &\leq u(c)\int_a^c \int_t^c |w(s)|dsdt \\ &= u(c)\int_a^c (s-a)|w(s)|ds \\ &\leq u(c)\int_a^b (s-a)|w(s)|ds \\ &\leq u(c)\int_a^b s|w(s)|ds, \end{aligned}$$

and the inequality is proved after canceling $u(c)$.

Remark 4.2. The proof is exactly the same as that in Chap. 2, due to Patula; see Lemma 2.1.

4.1.3 The Quasilinear Problem

The quasilinear problem is slightly more subtle, and we can obtain different inequalities with the same idea, depending on the cases $1 < p < 2$ and $p > 2$. Those inequalities improve Lyapunov's inequality in certain cases, and a full analysis of the interdependence has not yet been made.

We have the following theorem:

Theorem 4.2. *Let u be a nontrivial solution of*

$$-(|u'|^{p-2}u')' = w(x)|u|^{p-2}u, \qquad x \in (a,b),$$

satisfying $u(a) = u(b) = 0$, and w a positive weight in $L^1(a,b)$. Then if $1 < p < 2$, we have

$$1 \le \int_a^b (x-a)^{p-1}w(x)dx, \tag{1.6}$$

and for $p > 2$, we have

$$1 \le (b-a)^{p-2}\int_a^b (s-a)w(s)ds. \tag{1.7}$$

Proof. In much the same way, we can assume that u is positive in (a,c), and we take $c \in (a,b)$, where $|u|$ reaches its maximum, and we obtain

$$|u'(t)|^{p-2}u'(t) = \int_t^c w(s)|u|^{p-2}uds.$$

Now since $u(a) = 0$ and $u' > 0$ in (a,c), we get

$$
\begin{aligned}
u^{p-1}(c) &= \left(\int_a^c u'(t)dt\right)^{p-1} \\
&= \left(\int_a^c \left(\int_t^c w(s)|u(s)|^{p-2}u(s)ds\right)^{\frac{1}{p-1}} dt\right)^{p-1} \\
&\le u^{p-1}(c)\left(\int_a^c \left(\int_t^c w(s)ds\right)^{\frac{1}{p-1}} dt\right)^{p-1},
\end{aligned}
$$

which implies that

$$1 \le (b-a)^{p-2}\int_a^b (s-a)w(s)ds.$$

Observe that this bound is valid for every $1 < p < \infty$, so inequality (1.7) is proved.

We will now use Minkowski's integral inequality, which gives, for $r \geq 1$,

$$\left(\int_A \left| \int_B F(s,t) ds \right|^r dt \right)^{1/r} \leq \int_B \left(\int_A |F(s,t)|^r dt \right)^{1/r} ds.$$

We fix $A = B = [a,c]$, $r = 1/(p-1)$, and $F(x,t) = w(t)\chi_T(x,t)$, with χ_T the characteristic function of the triangle T,

$$T = \{(x,t) \in [a,c] \times [a,c] : x \leq t\}.$$

Thus,

$$\left(\int_a^c \left(\int_t^c w(s) ds \right)^{\frac{1}{p-1}} dt \right)^{p-1} \leq \int_a^c \left(\int_a^s w^{1/(p-1)}(s) dt \right)^{p-1} ds$$
$$\leq \int_a^c w(s)(s-a)^{p-1} ds,$$

and inequality (1.6) is proved. The proof is finished. $\square$

Remark 4.3. The previous theorem holds when $w \in L^1_{loc}(a,b)$, with $(x-a)^{p-1}w(x) \in L^1(a,b)$ for $1 < p < 2$, and $(x-a)w(x) \in L^1(a,b)$ for $p > 2$. We omit those easy steps in the proof.

Remark 4.4. Inequality (1.6) is better than inequality (1.7), since we can apply it to problems in unbounded intervals.

We wish to improve the result for $p > 2$, and in the next section we prove Theorems D and D', where a worse constant appears on the left-hand side.

4.2 Proofs of Theorems D and D'

The proof of Theorem D is based on a different idea, using the inequality of Chap. 3, and a similar argument to that in the proof of Lyapunov's inequality in Sect. 2.2.2.

Theorem 4.3. *Let λ_1 be the first eigenvalue of*

$$-u'' = \lambda w(x)u, \qquad x \in (a,b),$$

satisfying $u(a) = u(b) = 0$. Then

$$\frac{\pi}{4} \leq \lambda \int_a^b (x-a)w(x) dx. \tag{2.8}$$

Proof. We multiply the equation by u, and after integrating by parts, we get

$$\int_a^b u'^2 dx = \int_a^b \lambda w(x) u^2 dx.$$

Let us introduce the auxiliary function W,

$$W(x) = \int_x^b w(t)dt,$$

and thus, using integration by parts and Young's inequality,

$$\int_a^b u'^2 dx = \lambda \int_a^b w(x)u^2 dx$$

$$= -\lambda \int_a^b \frac{dW(x)}{dx} u^2 dx$$

$$= -\lambda W(x)u^2 \Big|_a^b + \lambda \int_a^b 2W(x)uu' dx$$

$$\leq 2 \int_a^b (\lambda W)^2 u^2 dx + \frac{1}{2} \int_a^b u'^2 p dx.$$

So after rearranging terms, we obtain

$$\int_a^b u'^2 dx \leq (2\lambda)^2 \int_a^b W^2(x)u^2 dx.$$

Let us consider now the eigenvalue problem

$$-v'' = \mu W^2 v, \qquad x \in (a,b),$$

with $v(a) = v(b) = 0$. The variational characterization of the first eigenvalue μ_1 shows that

$$\mu_1 = \inf_{u \in H_0^1(a,b)} \frac{\int_a^b u'^2 dx}{\int_a^b W^2 u^2} \leq (2\lambda_1)^2.$$

From Theorem C, we have the following lower bound for μ_1:

$$\frac{\pi}{2} \leq \mu_1^{\frac{1}{2}} \int_a^b W(x)dx \leq 2\lambda_1 \int_a^b W(x)dx.$$

Thus, by interchanging the order of integration, we obtain

$$\int_a^b \int_x^b w(t)dt dx = \int_a^b \int_a^t w(t)dx dt = \int_a^b (t-a)w(t)dt,$$

and the theorem is proved. □

Remark 4.5. The constant obtained is slightly worse than the previous one, since $1 > \pi/4$ when the angular moment is zero. However, the method of proof can be used in the quasilinear setting.

Theorem 4.4. *Let λ_1 be the first eigenvalue of*

$$-(|u'|^{p-2}u')' = \lambda w(x)|u|^{p-2}u, \qquad x \in (a,b),$$

satisfying $u(a) = u(b) = 0$, and w a positive weight in $L^1(a,b)$. Then we have

$$\frac{1}{p}\left(\frac{\pi_p}{2}\right)^{p-1} \le \lambda_1 \int_a^b (x-a)^{p-1}w(x)dx. \qquad (2.9)$$

Proof. We multiply the equation by u, and after integrating by parts, we get

$$\int_a^b |u'|^p dx = \int_a^b \lambda w(x)|u|^p dx.$$

We introduce the auxiliary function W,

$$W(x) = \int_x^b w(t)dt,$$

and thus, using integration by parts and Young's inequality, we obtain

$$\int_a^b |u'|^p dx = \lambda \int_a^b w(x)|u|^p dx$$

$$= -\lambda \int_a^b \frac{dW(x)}{dx}|u|^p dx$$

$$= -\lambda W(x)|u|^p \Big|_a^b + \lambda \int_a^b W(x)p|u|^{p-1}u'dx$$

$$\le \frac{p-1}{p}\int_a^b (\lambda pW)^{\frac{p}{p-1}}|u|^p dx + \frac{1}{p}\int_a^b |u'|^p dx.$$

Rearranging terms, we obtain

$$\int_a^b |u'|^p dx \le (\lambda p)^{\frac{p}{p-1}} \int_a^b W^{\frac{p}{p-1}}(x)|u|^p dx.$$

Let us consider now the eigenvalue problem

$$-(|v'|^{p-2}u')' = \mu W^{\frac{p}{p-1}}(x)|v|^{p-2}v, \qquad x \in (a,b),$$

with $u(a) = u(b) = 0$. The variational characterization of the first eigenvalue μ_1 shows that

$$\mu_1 = \inf_{u \in W_0^{1,p}(a,b)} \frac{\int_a^b |u'|^p dx}{\int_a^b W^{\frac{p}{p-1}}|u|^p} \le (p\lambda)^{\frac{p}{p-1}},$$

and this inequality is obtained when u_1, the associated eigenfunction to λ_1, is used as a test function.

Since w is positive, W is a nonincreasing function, and Theorem 3.5 with exponent $q = p' = p/(p-1)$ gives the following lower bound for μ_1 and therefore for λ_1:

$$\frac{1}{(p-1)\int_a^b (x-a)^{p-2}W(x)\mathrm{d}x}\left(\frac{\pi_p}{2}\right)^{p-1} \le \mu_1^{\frac{p-1}{p}} \le p\lambda_1. \qquad (2.10)$$

Equivalently,

$$\left(\frac{\pi_p}{2}\right)^{p-1} \le p(p-1)\lambda_1 \int_a^b (x-a)^{p-2}W(x)\mathrm{d}x.$$

Finally, using Fubini's theorem, we get

$$\int_a^b (x-a)^{p-2}\left(\int_x^b w(t)\mathrm{d}t\right)\mathrm{d}x = \int_a^b w(t)\left(\int_a^t (x-a)^{p-2}\mathrm{d}x\right)\mathrm{d}t$$
$$= \frac{1}{p-1}\int_a^b w(t)(t-a)^{p-1}\mathrm{d}t,$$

and replacing this in Eq. (2.10), we get

$$\left(\frac{\pi_p}{2}\right)^{p-1} \le p\lambda_1 \int_a^b w(t)(t-a)^{p-1}\mathrm{d}t.$$

The proof is finished. $\square$

Remark 4.6. We omit the proof of a dual inequality including a power of $(b-t)$ in the integrand,

$$\left(\frac{\pi_p}{2}\right)^{p-1} \le p\lambda_1 \int_a^b w(t)(b-t)^{p-1}\mathrm{d}t,$$

which can be proved in much the same way, using now $W(x) = \int_a^x w(t)\mathrm{d}t$.

4.3 Comparison of Inequalities

Let us compare now the different bounds obtained from Theorems A–D. We analyze only the linear case $p = 2$, since with easy modifications of the examples considered, we can consider any $p \in (1,\infty)$.

We fix the following Dirichlet eigenvalue problem in $(0,L)$:

$$-u'' = \lambda w(x)u, \qquad u(0) = u(L) = 0,$$

with $w \geq 0$, and we have

(A)   Sturm
$$\frac{\pi^2}{\max_{x \in (0,L)}\{w(x)\}L^2} \leq \lambda_w,$$

(B)   Lyapunov
$$\frac{4}{L \int_0^L w(x)dx} \leq \lambda_w,$$

(C)   Nehari–Calogero–Cohn
$$\left(\frac{\pi}{2 \int_0^L w^{\frac{1}{2}}(x)dx}\right)^2 \leq \lambda_w,$$

(D)   Bargmann
$$\frac{\pi}{4 \int_0^L xw(x)dx} \leq \lambda_w.$$

The bound in Sturm's theorem, Theorem A.9, is better when the weight is uniformly distributed, and it is sharp for a constant, $w(x) = C$, although in this case, Lyapunov's inequality is close, except by the factor 4 instead of π^2; Theorem C has a factor $1/4$, and Theorem D is the worst:

$$\frac{\pi}{2L^2C} < \frac{\pi^2}{4L^2C} < \frac{4}{L^2C} < \frac{\pi^2}{L^2C} = \lambda_w.$$

Let us observe that all the bounds include the correct powers of L and C, and only the constants differ.

Lyapunov's inequality is optimal when the weight is concentrated near the center of the interval. Suppose that $w(x) = \varepsilon^{-1}$ in $(L/2, L/2+\varepsilon)$, with ε positive and small, and zero outside. We have

$$\frac{\varepsilon\pi^2}{L^2} \ll \frac{\pi}{2(L+\varepsilon)} \leq \frac{4}{L} \approx \lambda_w,$$

where the Sturmian bound is the worse than the others, and we cannot apply Theorem C, since the weight is not monotonic.

The last two theorems are better for weights concentrated near the boundary. For example, $w(x) = \varepsilon^{-1}$ in $(0, \varepsilon)$ with ε positive and small, and zero in (ε, L). Since

$$(\|w\|_{L^\infty})^{-1} = \varepsilon, \quad \text{and} \quad \|w\|_{L^1} = 1,$$

the better bounds are given by Theorems C and D, since

$$\int_0^L xw(x) = \frac{\varepsilon}{2}, \quad \text{and} \quad \left(\int_0^L w^{\frac{1}{2}}(x)dx\right)^2 = \varepsilon,$$

and the eigenvalue is bounded below by $O(\varepsilon^{-1})$. Of course, after a little shift of the weight such that $w(x) = \varepsilon^{-1}$ in $(\varepsilon, 2\varepsilon)$ and zero outside, we cannot use Theorem C.

Indeed, Theorems C and D are useful for equations defined on the half-line, or when singular weights are involved. The case of unbounded intervals will be considered later, and let us mention briefly the singular weight $w(x) = x^{-1}$. This is an interesting case, which appears in the problem of rotating chains studied by Kolodner; see [71]. Clearly, we cannot apply the Sturmian theory or Lyapunov's inequality in $(0, L)$, and we can use Theorems C and D to find lower bounds of the eigenvalues of inhomogeneous chains that cannot be solved explicitly.

4.4 Singular Eigenvalue Problems

4.4.1 Introduction

In this section we review some linear and quasilinear singular eigenvalue problems, like those considered in [6, 44, 64, 69, 72, 84, 85, 96], and we analyze the asymptotic behavior of eigenvalues.

We are interested in the values of λ such that the second-order linear ordinary differential equation

$$u'' + \lambda w(x)u = 0, \quad x \geq a \tag{4.11}$$

with initial conditions

$$u(a, \lambda) = 0, \qquad u'(a, \lambda) = 1 \tag{4.12}$$

has a solution satisfying

$$\lim_{x \to \infty} u(x, \lambda) = c$$

for some real constant c.

It is well known that this problem can be classified as oscillatory or nonoscillatory depending on the number of zeros of $w(x, \lambda)$. We say that Eq. (4.11) is oscillatory if every solution has infinitely many zeros, and nonoscillatory if every solution has a finite number of zeros.

We will work with equations that are nonoscillatory for every $\lambda > 0$, and this is equivalent to imposing the following condition on w:

$$\int_a^\infty xw(x)dx < \infty. \tag{4.13}$$

We have the following existence result:

Theorem 4.5. *For every w satisfying condition (4.13), there exist a sequence of eigenvalues $\{\lambda_n\}_{n \geq 1}$ and associated solutions $\{u_n\}_{n \geq 1}$ of problem (4.11) with boundary conditions (4.12) such that*

$$\lim_{x \to \infty} u_n(x, \lambda_n) = c_n,$$

where c_n is a real constant.

Remark 4.7. We will write $\lambda_n^{(a)}$ and $u_n^{(a)}$ whenever we need to stress the dependence on the interval.

The proof of Theorem 4.5 goes back to Hille; see [64] and the references therein. For higher-order linear differential equations, see [44, 85], and [72] for quasilinear equations.

Following Hille, we can study the rate of growth of λ_n, and two different cases arise, depending on the convergence of the integral

$$I := \int_a^\infty w^{\frac{1}{2}}(x) dx.$$

We have the following results:

Theorem 4.6. *Let $\{\lambda_n\}_{n \geq 1}$ be the sequence of eigenvalues given by Theorem 4.5. Let w be a positive, continuous, and nonincreasing function satisfying $I < \infty$. Then*

$$\lambda_n = \left(\frac{\pi n}{I}\right)^2 + o(n^2).$$

Theorem 4.7. *Let $\{\lambda_n\}_{n \geq 1}$ be the sequence of eigenvalues given by Theorem 4.5. Let w be a positive, continuous, and nonincreasing function satisfying $I = \infty$. Then*

$$\frac{n-1}{\int_a^\infty xw(x) dx} \leq \lambda_n.$$

Moreover,

$$\frac{\lambda_n^{(a)}}{n^2} \to 0$$

as $n \to \infty$.

Remark 4.8. The first part follows easily from Bargmann's bound: the eigenfunction $u_n(x, \lambda_n)$ has exactly n zeros; let us call them $a = x_1 < x_2 < \cdots < x_n$. We can apply bound between two consecutive zeros $(n-1)$ times, obtaining

$$n - 1 \leq \lambda_n^{(a)} \left(\int_{x_1}^{x_2} xw(x) dx + \cdots + \int_{x_{n-1}}^{x_n} xw(x) dx\right) \leq \lambda_n^{(a)} \int_a^\infty xw(x) dx.$$

Remark 4.9. For some special weight functions, Hille obtained better estimates, which are particular cases of the estimate proved by Birman, Naimark, and Solomyak [6, 7, 84, 105],

$$\lambda_n = O(n^\alpha)$$

where $1 < \alpha < 2$ depends on the weight. See also Remark 4.16 for a different approach to obtaining examples of those estimates.

Remark 4.10. There are three different proofs of Theorem 4.6. Hille considered the Prüfer transformation method, which can be used for both linear and quasilinear problems of second order. Birman, Laptev, Naimark, and Solomyak used spaces of sequences of singular numbers of compact operators and an appropriate decomposition of Hilbert spaces together with difficult techniques from functional analysis; this method was applied to linear problems of all orders. Finally, in [96], we used Weyl's estimate for eigenvalues in finite intervals together with Theorem C; the extension to quasilinear and higher-order problems can be obtained in a similar fashion.

It is remarkable that all the proofs need the monotonicity of the weight despite the differences in the proofs: Hille needs to bound some integral; Birman, Laptev, and Solomyak constructed some associated decaying sequences; and in [96], this condition appears in the hypotheses of Theorem C.

4.4.2 The Asymptotic Behavior of Eigenvalues

4.4.2.1 Proof of Theorem 4.6

We give here an alternative formulation of Theorem 4.6. First, we introduce the spectral counting function $N(\lambda, \{\lambda_k\}_{k\geq 1})$, defined as

$$N(\lambda, \{\lambda_n\}_{n\geq 1}) = \#\{n : \lambda_n \leq \lambda\}.$$

Usually, we will omit the dependence on the sequence $\{\lambda_n\}_{n\geq 1}$.

Let us note that from an expression of the form

$$N(\lambda, \{\lambda_n\}_{n\geq 1}) = C\lambda^{\gamma} + o(\lambda^{\gamma}),$$

we get

$$n = N(\lambda_n) = C\lambda_n^{\gamma} + o(\lambda_n^{\gamma}),$$

and therefore

$$\lambda_n = cn^{\frac{1}{\gamma}} + o(n^{\frac{1}{\gamma}}).$$

This approach was pioneered by Weyl, and later studied by several mathematicians [31, 70]. Given a finite interval (a, b) and the sequence of eigenvalues $\{\lambda_n\}_{n\geq 1}$ corresponding to the second-order equation

$$-u'' = \lambda w(x)u \tag{4.14}$$

with Dirichlet or Neumann boundary conditions, the asymptotic expansion for $N(\lambda)$ is given by

$$N(\lambda) = \frac{\lambda^{\frac{1}{2}}}{\pi} \int_a^b w^{\frac{1}{2}}(x)dx + o(\lambda^{\frac{1}{2}})$$

as $\lambda \to \infty$. A similar formula holds for the p-Laplacian,

$$N(\lambda) = \frac{\lambda^{\frac{1}{p}}}{\pi_p} \int_a^b w^{\frac{1}{p}}(x)dx + o(\lambda^{\frac{1}{p}}).$$

We prove now the following theorem, which generalizes the Courant Dirichlet–Neumann bracketing (see [31, 48]):

Theorem 4.8. *Let $a < b < \infty$, and let $\{\lambda_n^{(a)}\}_{n\geq 1}$ and $\{\lambda_n^{(b)}\}_{n\geq 1}$ be the sequences given by Theorem 4.5 in (a,∞) and (b,∞) respectively. Let $\{\mu_n\}_{n\geq 1}$ be the sequence of eigenvalues of Eq. (4.14) in $[a,b]$ with Dirichlet boundary conditions. Then*

$$N(\lambda, \{\lambda_n^{(a)}\}_{n\geq 1}) \sim N(\lambda, \{\mu_n\}_{n\geq 1}) + N(\lambda, \{\lambda_n^{(b)}\}_{n\geq 1})$$

as $\lambda \to \infty$, where $f(\lambda) \sim g(\lambda)$ means that $f(\lambda)/g(\lambda) \to 1$ as $\lambda \to \infty$.

Proof. A key point in the following proof is the fact that the eigenfunction $u_n^{(a)}$ has exactly n simple zeros in $[a,\infty)$. The same is true for higher-order and quasilinear problems.

Let us take $\lambda \geq \lambda_1^{(a)}$. There exists some $n \geq 1$ such that

$$\lambda_n^{(a)} \leq \lambda < \lambda_{n+1}^{(a)},$$

and we have

$$N(\lambda, \{\lambda_n^{(a)}\}_{n\geq 1}) = n. \tag{4.15}$$

We can assume that k zeros of u_n belong to $[a,b)$ and the remaining $n-k$ zeros are located in $[b,\infty)$.

Sturm's comparison theorem, Theorem 1.3, implies that at least k zeros of u_{n+1} are contained in the interval $[a,b)$, and $n-k-1$ in $[b,\infty)$, and there are three possibilities: u_{n+1} has

- k zeros in $[a,b)$, $n-k+1$ in $[b,\infty)$;
- $k+1$ zeros in $[a,b)$, $n-k$ in $[b,\infty)$; and
- $k+2$ zeros in $[a,b)$, $n-k-1$ in $[b,\infty)$,

since the zeros travel to the left, and perhaps one of them crosses b.

We can consider now the solution u_λ of Eq. (4.14) satisfying

$$u_\lambda(a) = 0, \qquad u_\lambda'(a) = 1.$$

Again, Sturm's comparison theorem implies that u_λ has

- at least k zeros in $[a,b)$, and at most $k+2$;
- at least $n-k-1$ zeros in $[a,b)$, and at most $n-k+1$.

We are ready to bound $N(\lambda, \{\mu_n\}_{n\geq 1})$, the Dirichlet eigenvalues in $[a,b]$. Clearly, there exists some $j \geq 1$ such that

$$\mu_j \leq \lambda < \mu_{j+1},$$

and the Sturmian theory implies that u_λ has j zeros in $[a,b)$. So, from the arguments in the preceding paragraph we have $k \leq j \leq k+2$, which implies that

$$k \leq N(\lambda, \{\mu_n\}_{n\geq 1}) \leq k+2.$$

The bound for $N(\lambda, \{\lambda_n^{(b)}\}_{n\geq 1})$ is similar,

$$\lambda_i^{(b)} \leq \lambda < \lambda_{i+1}^{(b)},$$

for some $i \geq 1$, with $i \leq n-k+1$, and $n-k-1 \leq i+1$, which gives

$$n-k \leq N(\lambda, \{\lambda_n^{(b)}\}_{n\geq 1}) \leq n-k+1.$$

Thus,

$$N(\lambda, \{\mu_n\}_{n\geq 1}) + N(\lambda, \{\lambda_n^{(b)}\}_{n\geq 1}) \geq k+n-k-1$$
$$= n-1$$
$$= N(\lambda, \{\lambda_n^{(a)}\}_{n\geq 1}) - 1$$

$$N(\lambda, \{\mu_n\}_{n\geq 1}) + N(\lambda, \{\lambda_n^{(b)}\}_{n\geq 1}) \leq k+2+n-k+1$$
$$= n+3$$
$$= N(\lambda, \{\lambda_n^{(a)}\}_{n\geq 1}) + 3,$$

and the proof is finished.   □

We can prove now the following results:

Theorem 4.9. *Let $\{\lambda_n^{(a)}\}_{n\geq 1}$ be the sequence of eigenvalues given by Theorem 4.5. Let w be a positive continuous function satisfying $I < \infty$. Then for every $\varepsilon > 0$, there exists some Λ_0 such that*

$$\frac{\lambda^{\frac{1}{2}}}{\pi} \int_a^{+\infty} w^{\frac{1}{2}}(x)dx - \varepsilon\lambda^{\frac{1}{2}} \leq N(\lambda, \{\lambda_n^{(a)}\}_{n\geq 1}) \tag{4.16}$$

for every $\lambda > \Lambda_0$. Equivalently,

$$\lambda_n^{(a)} \leq \left(\frac{\pi n}{I}\right)^2 + c_\varepsilon n^2$$

with $c_\varepsilon \to 0^+$ when $\varepsilon \to 0$, and $\lambda_n > \Lambda_0$.

Proof. Let $\varepsilon > 0$ be fixed, and let us choose T_ε such that

$$\frac{1}{\pi} \int_{T_\varepsilon}^\infty w^{\frac{1}{2}}(x)dx \leq \frac{\varepsilon}{2}.$$

Let $\{\mu_n\}_{n\geq 1}$ be the sequence of eigenvalues of Eq. (4.14) in $[a,b]$ with Dirichlet boundary conditions. Since $\lambda_n \leq \mu_n$ (we are using Sturm's theorem), we have

$$N(\lambda, \{\mu_n\}_{n\geq 1}) \leq N(\lambda, \{\lambda_n^{(a)}\}_{n\geq 1}).$$

Since

$$N(\lambda, \{\mu_n\}_{n\geq 1}) = \frac{\lambda^{\frac{1}{2}}}{\pi} \int_a^{T_\varepsilon} w^{\frac{1}{2}}(x)dx + o(\lambda^{\frac{1}{2}}), \tag{4.17}$$

there exists λ_0 such that

$$\left| N(\lambda, \{\mu_n\}_{n\geq 1}) - \frac{\lambda^{\frac{1}{2}}}{\pi} \int_a^{T_\varepsilon} w^{\frac{1}{2}}(x)dx \right| \leq \frac{\varepsilon}{2}\lambda^{\frac{1}{2}}$$

whenever $\lambda \geq \Lambda_0$. Thus

$$\begin{aligned}
N(\lambda, \{\lambda_n^{(a)}\}_{n\geq 1}) &\geq N(\lambda, \{\mu_n\}_{n\geq 1}) \\
&\geq \frac{\lambda^{\frac{1}{2}}}{\pi} \int_a^{T_\varepsilon} w^{\frac{1}{2}}(x)dx - \frac{\varepsilon}{2}\lambda \\
&\geq \frac{\lambda^{\frac{1}{2}}}{\pi} \int_a^{\infty} w^{\frac{1}{2}}(x)dx - \frac{\lambda^{\frac{1}{2}}}{\pi} \int_{T_\varepsilon}^{\infty} w^{\frac{1}{2}}(x)dx - \frac{\varepsilon}{2}\lambda^{\frac{1}{2}} \\
&\geq \frac{\lambda^{\frac{1}{2}}}{\pi} \int_a^{\infty} w^{\frac{1}{2}}(x)dx - \varepsilon\lambda^{\frac{1}{2}}.
\end{aligned}$$

The bound for $\lambda_n^{(a)}$ follows by replacing it in the previous inequality, and the theorem is proved. $\square$

Our next result gives an upper bound for the spectral counting function:

Theorem 4.10. *Let* $\{\lambda_n^{(a)}\}_{n\geq 1}$ *be the sequence of eigenvalues given by Theorem 4.5. Let w be a positive, continuous, and nonincreasing function satisfying $I < \infty$. Then for every $\varepsilon > 0$, there exists some Λ_0 such that*

$$N(\lambda, \{\lambda_n^{(a)}\}_{n\geq 1}) \leq \frac{\lambda^{\frac{1}{2}}}{\pi} \int_a^{+\infty} w^{\frac{1}{2}}(x)dx + \varepsilon\lambda^{\frac{1}{2}} \tag{4.18}$$

for every $\lambda > \Lambda_0$. Equivalently,

$$\lambda_n \leq \left(\frac{\pi n}{I}\right)^2 + c_\varepsilon n^2,$$

with $c_\varepsilon \to 0^+$ when $\varepsilon \to 0$, and $\lambda_n > \Lambda_0$.

Proof. For fixed $\varepsilon > 0$, we can choose $b > a$ depending on ε such that

$$\frac{6}{\pi^2} \int_b^{\infty} w^{\frac{1}{2}}(x)dx \leq \varepsilon.$$

Hence, from Theorem 5.27, specializing $p = 2$ and extending the integral beyond the last zero of the eigenfunction $u_n^{(b)}$, we have the bound

$$\frac{n^2\pi^2}{4} < \lambda_n^{(b)} \left(\int_b^\infty w^{\frac{1}{2}}(x) dx \right)^2 .$$

Thus,

$$N(\lambda, \{\lambda_n^{(b)}\}_{n\geq 1}) \leq \# \left\{ n : \frac{n^2\pi^2}{4(\int_b^\infty w^{\frac{1}{2}}(x)dx)^2} \leq \lambda \right\}$$

$$\leq \frac{2\lambda^{\frac{1}{2}}}{\pi} \int_b^\infty w^{\frac{1}{2}}(x) dx \tag{4.19}$$

$$\leq \frac{\varepsilon\lambda^{\frac{1}{2}}}{3} .$$

On the other hand,

$$N(\lambda, \{\mu_n\}_{n\geq 1}) \leq \frac{\lambda^{\frac{1}{2}}}{\pi} \int_a^b w^{\frac{1}{2}}(x) dx + \frac{\varepsilon\lambda^{\frac{1}{2}}}{3}, \tag{4.20}$$

for every $\lambda > \Lambda_1$, where $\{\mu_n\}_{n\geq 1}$ are the eigenvalues of Eq. (4.14) in $[a,b]$ with Dirichlet boundary conditions.

We now use Theorem 4.8. For fixed ε, we can choose δ_ε such that

$$N(\lambda, \{\lambda_n^{(a)}\}_{n\geq 1}) \leq (1 + \delta_\varepsilon) \left(N(\lambda, \{\mu_n\}_{n\geq 1}) + N(\lambda, \{\lambda_n^{(b)}\}_{n\geq 1}) \right) \tag{4.21}$$

for every $\lambda > \Lambda_2$, where Λ_2 depends on δ_ε..

Finally, we can replace the bounds (4.19) and (4.20) in Eq. (4.21), and clearly we can choose δ_ε such that

$$N(\lambda, \{\lambda_n^{(a)}\}_{n\geq 1}) \leq \frac{\lambda^{\frac{1}{2}}}{\pi} \int_a^{+\infty} w^{\frac{1}{2}}(x) dx + \varepsilon\lambda^{\frac{1}{2}}$$

for every $\lambda > \Lambda_0 = \max\{\Lambda_1, \Lambda_2\}$.

The proof is now complete.   □

The following result is a corollary of Theorems 4.9 and 4.10:

Theorem 4.11. *Let* $\{\lambda_n^{(a)}\}_{n\geq 1}$ *be the sequence of eigenvalues given by Theorem 4.5. Let w be a positive, continuous, and nonincreasing function satisfying $I < \infty$. Then*

$$N(\lambda, \{\lambda_n^{(a)}\}_{n\geq 1}) = \frac{\lambda^{\frac{1}{2}}}{\pi} \int_a^{+\infty} w^{\frac{1}{2}}(x) dx + o(\lambda^{\frac{1}{2}}) \tag{4.22}$$

as $\lambda \to \infty$.

The result follows, since we can take ε arbitrarily small in the bound

$$\left| N(\lambda, \{\lambda_n^{(a)}\}_{n\geq 1}) - \frac{\lambda^{\frac{1}{2}}}{\pi} \int_a^{+\infty} w^{\frac{1}{2}}(x)\mathrm{d}x \right| \leq \varepsilon\lambda^{\frac{1}{2}}.$$

Remark 4.11. Let us observe that Theorem 4.6 follows directly, since

$$n = N(\lambda, \{\lambda_n^{(a)}\}_{n\geq 1}) = (\lambda_n^{(a)})^{\frac{1}{2}}(1+o(1))\left(\frac{1}{\pi}\int_a^{+\infty} w^{\frac{1}{2}}(x)\mathrm{d}x\right),$$

and then

$$\lambda_n^{(a)} = (1+o(1))^{-2}\left(\frac{n\pi}{\int_a^\infty w^{\frac{1}{2}}(x)\mathrm{d}x}\right)^2. \tag{4.23}$$

Remark 4.12. When Eq. (4.22) holds, we say that $\{\lambda_n^{(a)}\}_{n\geq 1}$ satisfy a Weyl-type asymptotic. So we have proved that the eigenvalues given by Theorem 4.5 satisfy a Weyl-type asymptotic for monotonic weights w when $I < \infty$.

Remark 4.13. The result of this section can be proved also for quasilinear problems. Higher-order problems are more difficult, since we cannot use the Sturm–Liouville oscillation theory. We must rely on the existence of conjugate points and extremal solutions.

4.4.3 Proof of Theorem 4.7

In this section we consider the case $I = \infty$. Let us observe first that Theorem 4.8 is valid in this case, and then we have

$$N(\lambda, \{\lambda_n^{(a)}\}_{n\geq 1}) \sim N(\lambda, \{\mu_n\}_{n\geq 1}) + N(\lambda, \{\lambda_n^{(b)}\}_{n\geq 1}) \tag{4.24}$$

as $\lambda \to \infty$ for every $b \in (a,\infty)$.

Moreover, there are no monotonicity assumptions on w, and we can prove the following result analogous to Theorem 4.9:

Theorem 4.12. *Let $\{\lambda_n^{(a)}\}_{n\geq 1}$ be the sequence of eigenvalues given by Theorem 4.5. Let w be a positive continuous function. Then for fixed $b > a$, there exists Λ_0 such that*

$$N(\lambda, \{\lambda_n^{(a)}\}_{n\geq 1}) \geq \frac{\lambda^{\frac{1}{2}}}{\pi}\int_a^b w^{\frac{1}{2}}(x)\mathrm{d}x + o(\lambda^{\frac{1}{2}}) \tag{4.25}$$

for every $\lambda \geq \Lambda_0$. Equivalently,

$$\lambda_n^{(a)} \leq \left(\frac{\pi n}{\int_a^b w^{\frac{1}{2}}(x)\mathrm{d}x}\right)^2 + o(n^2).$$

Proof. Let $\{\mu_n\}_{n\geq 1}$ be the sequence of eigenvalues of Eq. (4.14) in $[a,b]$ with Dirichlet boundary conditions. Sturm's theorem implies that $\lambda_n \leq \mu_n$, and we have

$$N(\lambda, \{\mu_n\}_{n\geq 1}) \leq N(\lambda, \{\lambda_n^{(a)}\}_{n\geq 1}).$$

The result follows from the fact that

$$N(\lambda, \{\mu_n\}_{n\geq 1}) = \frac{\lambda^{\frac{1}{2}}}{\pi} \int_a^b w^{\frac{1}{2}}(x)dx + o(\lambda^{\frac{1}{2}}).$$

$\square$

We can obtain now the proof of the upper bound for $\lambda_n^{(a)}$ in Theorem 4.7: we have

$$\frac{\lambda_n^{(a)}}{n^2} \leq \frac{1}{n^2}\left(\frac{\pi n}{\int_a^b w^{\frac{1}{2}}(x)dx}\right)^2 + o(1)$$

$$= \left(\frac{\pi}{\int_a^b w^{\frac{1}{2}}(x)dx}\right)^2 + o(1),$$

and this expression is arbitrarily small by choosing b big enough, since $I = \infty$.

Remark 4.14. It is known that both bounds in Theorem 4.7 are sharp, and there are examples of weights such that $\lambda_n^{(a)} = O(n^\alpha)$ for $1 < \alpha < 2$; see [64]. The case $\lambda_n^{(a)} = O(n)$ can be found in [44]; in fact, the eigenvalues and eigenfunctions of the differential equation

$$-u'' = \frac{\lambda}{x^2 \log(x)}u, \quad 1 < x < \infty$$

can be computed explicitly, $\lambda_n = n$, and

$$u_{\lambda_n}(x) = \log(x)L_{n-1}^{(1)}(\log(x)),$$

where $L_m^{(a)}$ denotes the Laguerre polynomials.

Remark 4.15. From the inequality

$$\frac{n-1}{\int_a^\infty xw(x)dx} \leq \lambda_n$$

in Theorem 4.7, we get

$$N(\lambda, \{\lambda_n^{(b)}\}_{n\geq 1}) \leq \lambda \int_b^\infty xw(x)dx.$$

Hence, from Eq. (4.24), we have the following inequality:

$$N(\lambda, \{\lambda_n^{(a)}\}_{n\geq 1}) \leq N(\lambda, \{\mu_n\}_{n\geq 1}) + \lambda \int_b^\infty xw(x)dx.$$

Moreover, given w a monotonic weight, we can use Theorem 5.27, which gives

$$\frac{n^2\pi^2}{4} < \lambda_n \left(\int_a^b w^{\frac{1}{2}}(x)dx \right)^2,$$

in order to bound $N(\lambda, \{\mu_n\}_{n\geq 1})$, obtaining

$$N(\lambda, \{\lambda_n^{(a)}\}_{n\geq 1}) \leq \frac{2\lambda^{\frac{1}{2}}}{\pi} \int_a^b w^{\frac{1}{2}}(x)dx + \lambda \int_b^\infty xw(x)dx.$$

Now we can choose $b = b(\lambda)$, and Bolzano's theorem implies that there exists some $b^* = b^*(\lambda)$ such that

$$\frac{2\lambda^{\frac{1}{2}}}{\pi} \int_a^{b^*} w^{\frac{1}{2}}(x)dx = \lambda \int_{b^*}^\infty xw(x)dx.$$

Therefore, we have the following asymptotic estimate:

$$N(\lambda, \{\lambda_n^{(a)}\}_{n\geq 1}) = O\left(\lambda \int_{b^*(\lambda)}^\infty xw(x)dx \right). \qquad (4.26)$$

Remark 4.16. The estimate (4.26) enables us characterize the asymptotic behavior of the eigenvalues in several cases. As an example, let us choose the family of functions

$$w(x) = \left(x^2 \log^\alpha(x) \right)^{-1},$$

with $1 < \alpha < 2$, and let $a = e$.

We have

$$\int_e^\infty w^{\frac{1}{2}}(x) = \int_e^\infty \frac{dx}{x\log^{\frac{\alpha}{2}}(x)} = \infty,$$

$$\int_e^\infty xw(x)dx = \int_e^\infty \frac{dx}{x\log^\alpha(x)} < \infty.$$

Moreover, by choosing $b = e^{\lambda^{1/\alpha}}$, we obtain

$$\int_{q_\infty}^b w^{\frac{1}{2}}(x)dx = O(\lambda^{\frac{1}{\alpha}-\frac{1}{2}}),$$

$$\int_b^\infty xw(x)dx = O(\lambda^{\frac{1}{\alpha}-1}),$$

and the estimate (4.26) implies that

$$N(\lambda, \{\lambda_n^{(a)}\}_{n\geq 1}) = O\left(\lambda^{\frac{1}{\alpha}} \right),$$

or equivalently,

$$\lambda_n = O(n^\alpha).$$

Chapter 5
Miscellaneous Topics

Abstract In this chapter we prove several Lyapunov-type inequalities for systems of ordinary differential equations, one-dimensional nonlinear operators in Orlicz spaces, and quasilinear equations in R^N.

5.1 Resonant Systems

The results in this section are based on joint work with Pablo de Napoli [36] and Julián Fernández Bonder [48], and several recent articles generalizing those results; see, for instance, [15, 16, 108, 115].

Let (u,v) be the eigenpair associated to the first eigenvalue λ_1 of the following system,

$$\begin{cases} -(|u'|^{p-2}u')' = \lambda w(x)\alpha|u|^{\alpha-2}u|v|^{\beta} & x \in (a,b), \\ -(|v'|^{q-2}v')' = \lambda w(x)\beta|u|^{\alpha}|v|^{\beta-2}v & x \in (a,b), \\ u(a) = u(b) = v(a) = v(b) = 0, \end{cases} \tag{1.1}$$

and let us recall that $w \in L^1(a,b)$ is allowed to change sign, and

$$\frac{\alpha}{p} + \frac{\beta}{q} = 1, \qquad 1 < q \leq p < \infty. \tag{1.2}$$

There are few results on the nodal structure of the eigenfuctions of system (1.1), and let us note that if (u,v) is the first eigenpair, i.e.,

$$\lambda_1 = \frac{\frac{1}{p}\int_a^b |u'|^p + \frac{1}{q}\int_a^b |v'|^q}{\int_a^b r(x)|u|^{\alpha}|v|^{\beta}} = \frac{\frac{1}{p}\int_a^b |u'|^p + \frac{1}{q}\int_a^b |v'|^q}{\int_a^b r(x)|u|^{\alpha}|v|^{\beta}}$$

then $(|u|,|v|)$ is also an eigenpair associated to λ_1. However, the uniqueness of solutions of the ordinary differential equation

J.P. Pinasco, *Lyapunov-type Inequalities: With Applications to Eigenvalue Problems*, SpringerBriefs in Mathematics, DOI 10.1007/978-1-4614-8523-0_5, © Juan Pablo Pinasco 2013

$$-(|u'|^{p-2}u')' = \lambda w(x)\alpha|u|^{\alpha-2}u|v|^{\beta}, \qquad u(a) = u(b) = 0,$$

implies that $u = \pm|u|$. So we have two pairs of eigenfunctions, $(|u|,\pm|v|)$. To our knowledge, there are no results about the number of nodal domains of the eigenpairs corresponding to higher eigenvalues.

5.1.1 Lyapunov's Inequality for Resonant Systems

The main theorem is the following:

Theorem 5.1. *Let us assume that there exists a pair (u,v) of positive solutions of system (1.1) on the interval (a,b), and suppose that $w \in L^1(a,b)$. Then we have*

$$\frac{2^{\alpha+\beta}}{\alpha^{\frac{\alpha}{p}}\beta^{\frac{\beta}{q}}} \leq (b-a)^{\alpha+\beta-1}\int_a^b w^+(x)\,dx, \tag{1.3}$$

where $w^+ = \max\{w(x),0\}$, as usual.

Proof. Let us call the point where $|u(x)|$ reaches its maximum c, and let d denote the point where $|v(x)|$ reaches its maximum. Hence

$$2|u(c)| = \left|\int_a^c u'(x)dx\right| + \left|\int_c^b u'(x)dx\right|$$

$$\leq \int_a^b |u'(x)|dx$$

$$\leq (b-a)^{\frac{p-1}{p}}\left(\int_a^b |u'(x)|^p dx\right)^{\frac{1}{p}}$$

$$= (b-a)^{\frac{p-1}{p}}\left(\int_a^b \alpha w(x)|u|^{\alpha}|v|^{\beta} dx\right)^{\frac{1}{p}}$$

$$\leq (b-a)^{\frac{p-1}{p}}\left(\alpha\int_a^b w^+(x)|u|^{\alpha}|v|^{\beta} dx\right)^{\frac{1}{p}},$$

$$2|v(d)| = \left|\int_a^d v'(x)dx\right| + \left|\int_d^b v'(x)dx\right|$$

$$\leq \int_a^b |v'(x)|dx$$

$$\leq (b-a)^{\frac{q-1}{q}}\left(\int_a^b |v'(x)|^p dx\right)^{\frac{1}{q}}$$

$$= (b-a)^{\frac{q-1}{q}}\left(\int_a^b \beta w(x)|u|^{\alpha}|v|^{\beta} dx\right)^{\frac{1}{q}}$$

$$\leq (b-a)^{\frac{q-1}{q}}\left(\beta\int_a^b w^+(x)|u|^{\alpha}|v|^{\beta} dx\right)^{\frac{1}{q}},$$

and using that $|u(x)| \leq |u(c)|$ and $|v(x)| \leq |v(d)|$, we get

$$2|u(c)| \leq (b-a)^{\frac{p-1}{p}}|u(c)|^{\frac{\alpha}{p}}|v(d)|^{\frac{\beta}{p}}\left(\alpha\int_a^b w^+(x)\,dx\right)^{\frac{1}{p}}, \tag{1.4}$$

$$2|v(d)| \leq (b-a)^{\frac{q-1}{q}}|u(c)|^{\frac{\alpha}{q}}|v(d)|^{\frac{\beta}{q}}\left(\beta\int_a^b w^+(x)\,dx\right)^{\frac{1}{q}}. \tag{1.5}$$

By raising Eq. (1.4) [resp., Eq. (1.5)] to the power α (resp., β), and after rearranging terms, we obtain

$$2^\alpha \leq (b-a)^{\frac{\alpha(p-1)}{p}}|u(c)|^{\alpha\left(\frac{\alpha}{p}-1\right)}|v(d)|^{\frac{\alpha\beta}{p}}\left(\alpha\int_a^b w^+(x)\,dx\right)^{\frac{\alpha}{p}},$$

$$2^\beta \leq (b-a)^{\frac{\beta(q-1)}{q}}|u(c)|^{\frac{\alpha\beta}{q}}|v(d)|^{\beta\left(\frac{\beta}{q}-1\right)}\left(\beta\int_a^b w^+(x)\,dx\right)^{\frac{\beta}{q}}.$$

Now multiplying both equations and using condition (1.2), we get

$$2^{\alpha+\beta} \leq (b-a)^{\alpha+\beta-1}\alpha^{\frac{\alpha}{p}}\beta^{\frac{\beta}{q}}\int_a^b w^+(x)\,dx,$$

since

$$|u(c)|^{\alpha\left(\frac{\alpha}{p}-1\right)}\cdot|u(c)|^{\frac{\alpha\beta}{q}} = |u(c)|^{\alpha\left(\frac{\alpha}{p}-1+\frac{\beta}{q}\right)} = 1$$

and

$$|v(d)|^{\frac{\alpha\beta}{p}}\cdot|v(d)|^{\beta\left(\frac{\beta}{q}-1\right)} = |v(d)|^{\beta\left(\frac{\alpha}{p}+\frac{\beta}{q}-1\right)} = 1.$$

The proof is finished. □

5.1.2 Some Generalizations

Theorem 5.1 can be generalized to systems without variational structure, provided that each equation can be integrated by parts and that they have some degree of homogeneity. For instance, for

$$-(|u'|^{p-2}u')' = f(x)\alpha|u|^{\alpha-2}u|v|^\beta,$$

$$-(|v'|^{q-2}v')' = g(x)\beta|u|^\alpha|v|^{\beta-2}v,$$

with Dirichlet boundary conditions $u(a) = u(b) = v(a) = v(b) = 0$, we can prove a similar inequality,

$$2^{\alpha+\beta} \le (b-a)^{\frac{\alpha}{p'}+\frac{\beta}{q'}} \left(\int_a^b f^+(x)\,dx \right)^{\frac{\alpha}{p}} \left(\int_a^b g^+(x)\,dx \right)^{\frac{\beta}{q}}.$$

We omit the proof, since it is identical to the previous one.

Also, several authors have considered quasilinear systems with more than two equations,

$$-(|u_i'|^{p-2}u_i')' = w_i(x)\alpha_i|u_i|^{\alpha_i-2}u_i \prod_{j\ne i}|u_j|^{\alpha_j}, \qquad 1 \le i \le k, \tag{1.6}$$

satisfying a zero Dirichlet boundary condition, $u_i(a) = u_i(b) = 0$, $w_i \in L^1(a,b)$, and the following homogeneity condition:

$$\sum_{i=1}^k \frac{\alpha_i}{p_i} = 1.$$

Let us set $\alpha = \sum_{i=1}^k \alpha_i$, and we have

$$\frac{2^\alpha}{(b-a)^{\alpha-1}} \le \prod_{i=1}^k \int_a^b w_i(x)\,dx.$$

The proof follows the same steps as before; see, for example, [16].

5.1.3 A Different Idea

An interesting inequality based on a different idea is due to Tang and He in [108]. They considered the system

$$-(r(x)|u'|^{p-2}u')' = f(x)\alpha|u|^{\alpha_1-2}u|v|^{\beta_1},$$

$$-(s(x)|v'|^{q-2}v')' = g(x)\beta|u|^{\alpha_2}|v|^{\beta_2-2}v,$$

with zero Dirichlet boundary conditions, both pairs (α_i,β_i) satisfying condition (1.2). We assume here for brevity that $r(x) \equiv s(x) \equiv 1$, $\alpha_1 = \alpha_2$, and $\beta_1 = \beta_2$, and that f and g are positive functions. The general case follows by Hölder's inequality, and f^+, g^+ as usual.

Theorem 5.2. *Let us assume that there exists a pair (u,v) of positive solutions of system (1.1) on the interval (a,b), and suppose that $w \in L^1(a,b)$ is a positive function. Then we have*

$$1 \leq \left(\int_a^b f(t)P(t)dt \right)^{\frac{\alpha^2}{p^2}} \cdot \left(\int_a^b g(t)P(t)dt \right)^{\frac{\alpha\beta}{pq}}$$

$$\cdot \left(\int_a^b f(t)Q(t)dt \right)^{\frac{\alpha\beta}{pq}} \cdot \left(\int_a^b g(t)Q(t)dt \right)^{\frac{\beta^2}{q^2}},$$

where

$$P(t) = \frac{[(b-t)(t-a)]^{p-1}}{(t-a)^{p-1} + (b-t)^{p-1}},$$

$$Q(t) = \frac{[(b-t)(t-a)]^{q-1}}{(t-a)^{q-1} + (b-t)^{q-1}}.$$

Proof. Starting from

$$|u(t)|^p \leq (t-a)^{p-1} \int_a^t |u'|^p dx, \qquad |u(t)|^p \leq (b-t)t^{p-1} \int_t^b |u'|^p dx,$$

$$|v(t)|^q \leq (t-a)^{q-1} \int_a^t |v'|^q dx, \qquad |v(t)|^q \leq (b-t)t^{q-1} \int_t^b |v'|^q dx,$$

by adding the inequalities we obtain

$$|u(t)|^p \leq \frac{[(b-t)(t-a)]^{p-1}}{(t-a)^{p-1} + (b-t)^{p-1}} \int_a^b |u'|^p dx = P(t) \int_a^b |u'|^p dx,$$

$$|v(t)|^q \leq \frac{[(b-t)(t-a)]^{q-1}}{(t-a)^{q-1} + (b-t)^{q-1}} \int_a^b |v'|^q dx = Q(t) \int_a^b |v'|^q dx.$$

Multiplying both sides of the equations first by f and then by g, we get

$$f(t)|u(t)|^p \leq f(t)P(t) \int_a^b |u'|^p dx, \tag{1.7}$$

$$g(t)|u(t)|^p \leq g(t)P(t) \int_a^b |u'|^p dx, \tag{1.8}$$

$$f(t)|v(t)|^q \leq f(t)Q(t) \int_a^b |v'|^q dx. \tag{1.9}$$

$$g(t)|v(t)|^q \leq g(t)Q(t) \int_a^b |v'|^q dx. \tag{1.10}$$

Next, we will use that

$$\int_a^b |u'|^p dx = \int_a^b f(x)|u|^\alpha |v|^\beta dx \leq \left(\int_a^b f(x)|u|^p dx \right)^{\frac{\alpha}{p}} \left(\int_a^b f(x)|v|^q dx \right)^{\frac{\beta}{q}},$$

$$\int_a^b |v'|^q \mathrm{d}x = \int_a^b g(x)|u|^\alpha |v|^\beta \mathrm{d}x \leq \left(\int_a^b g(x)|u|^p \mathrm{d}x \right)^{\frac{\alpha}{p}} \left(\int_a^b g(x)|v|^q \mathrm{d}x \right)^{\frac{\beta}{q}},$$

where we have used Hölder's inequality and condition (1.2).

Integrating from a to b on both sides of Eqs. (1.7)–(1.10), and using the previous inequalities, we obtain

$$\int_a^b f(t)|u(t)|^p \mathrm{d}t \leq \left(\int_a^b f(t)P(t)\mathrm{d}t \right) \tag{1.11}$$

$$\cdot \left(\int_a^b f(x)|u|^p \mathrm{d}x \right)^{\frac{\alpha}{p}} \cdot \left(\int_a^b f(x)|v|^q \mathrm{d}x \right)^{\frac{\beta}{q}},$$

$$\int_a^b g(t)|u(t)|^p \mathrm{d}t \leq \left(\int_a^b g(t)P(t)\mathrm{d}t \right) \tag{1.12}$$

$$\cdot \left(\int_a^b f(x)|u|^p \mathrm{d}x \right)^{\frac{\alpha}{p}} \cdot \left(\int_a^b f(x)|v|^q \mathrm{d}x \right)^{\frac{\beta}{q}},$$

$$\int_a^b f(t)|v(t)|^q \mathrm{d}t \leq \left(\int_a^b f(t)Q(t)\mathrm{d}t \right) \tag{1.13}$$

$$\cdot \left(\int_a^b g(x)|u|^p \mathrm{d}x \right)^{\frac{\alpha}{p}} \cdot \left(\int_a^b g(x)|v|^q \mathrm{d}x \right)^{\frac{\beta}{q}},$$

$$\int_a^b g(t)|v(t)|^q \mathrm{d}t \leq \left(\int_a^b g(t)Q(t)\mathrm{d}t \right) \tag{1.14}$$

$$\cdot \left(\int_a^b g(x)|u|^p \mathrm{d}x \right)^{\frac{\alpha}{p}} \cdot \left(\int_a^b g(x)|v|^q \mathrm{d}x \right)^{\frac{\beta}{q}}.$$

We raise both members of Eqs. (1.11)–(1.14) to different powers:

- Equation (1.11) to the power α^2/p^2,
- Equation (1.12) to the power $\alpha\beta/pq$,
- Equation (1.13) to the power $\alpha\beta/pq$,
- Equation (1.14) to the power β^2/q^2,

and after multiplying the members on each side of the inequalities, we get the desired result,

$$1 \leq \left(\int_a^b f(t)P(t)\mathrm{d}t \right)^{\frac{\alpha^2}{p^2}} \cdot \left(\int_a^b g(t)P(t)\mathrm{d}t \right)^{\frac{\alpha\beta}{pq}}$$

$$\cdot \left(\int_a^b f(t)Q(t)\mathrm{d}t \right)^{\frac{\alpha\beta}{pq}} \cdot \left(\int_a^b g(t)Q(t)\mathrm{d}t \right)^{\frac{\beta^2}{q^2}},$$

since the terms on the left-hand side cancel those on the right-hand side. For example,

$$\left(\int_a^b f(t)|u(t)|^p dt\right)^{\frac{\alpha^2}{p^2}} = \left(\int_a^b f(x)|u|^p dx\right)^{\frac{\alpha}{p}\frac{\alpha^2}{p^2}} \cdot \left(\int_a^b f(x)|u|^p dx\right)^{\frac{\alpha}{p}\frac{\alpha\beta}{pq}},$$

because

$$\frac{\alpha}{p}\frac{\alpha^2}{p^2} + \frac{\alpha}{p}\frac{\alpha\beta}{pq} = \frac{\alpha^2}{p^2}\left(\frac{\alpha}{p} + \frac{\beta}{q}\right) = \frac{\alpha^2}{p^2},$$

due to condition (1.2); the remaining terms are similar. □

5.1.4 Other Systems of Equations

There are other interesting nonlinear systems in the literature that are generalizations of a variety of linear systems. Indeed, we can write them as

$$-(|u_i'|^{p_i-2}u_i')' = F_i(u_1,\ldots,u_k), \qquad 1 \le i \le k,$$

imposing some homogeneity condition on the functions F_i. We present some particular examples.

5.1.4.1 Cycled Systems

This class mimics the systems obtained by reducing a higher-order differential equation to a system of second-order differential equations,

$$-(|u_1'|^{p-2}u_1')' = \lambda w_1(x)|u_2|^{p-2}u_2 \qquad x \in (a,b),$$
$$-(|u_2'|^{p-2}u_2')' = \lambda w_2(x)|u_3|^{p-2}u_3 \qquad x \in (a,b),$$

$$\ldots$$

$$-(|u_{k-1}'|^{p-2}u_{k-1}')' = \lambda w_{k-1}(x)|u_k|^{p-2}u_k \qquad x \in (a,b),$$
$$-(|u_k'|^{p-2}u_k')' = \lambda w_k(x)|u_1|^{p-2}u_1 \qquad x \in (a,b),$$

with zero Dirichlet boundary conditions,

$$u_i(a) = u_i(b) = 0, \qquad 1 \le i \le k.$$

5.1.4.2 Vector p-Laplacian Systems

Let us denote the vectors

$$U(x) = \begin{pmatrix} |u_1|^{p-2}u_1 \\ \cdots \\ |u_k|^{p-2}u_k \end{pmatrix}, \qquad DU(x) = \begin{pmatrix} |u_1'|^{p-2}u_1' \\ \cdots \\ |u_k'|^{p-2}u_k' \end{pmatrix}$$

by $U(x)$, $DU(x)$, where $u_i \in W_0^{1,p}(a,b)$ for $1 \leq i \leq k$. Given $W(x) \in [L^1(a,b)]^{k \times k}$, we have the problem

$$DU' = W(x)U(x), \qquad x \in (a,b),$$

with $U(a) = U(b) = 0$.

For linear systems of this type, the problem was studied by Reid [100], and some new results were proved recently by Cañada and Villegas in [20].

5.1.4.3 Coupled Differential Operators of Different Orders

This kind of system was introduced by Lazer and McKenna [74], and such systems appear in models of suspension bridges. A simple example reads

$$-u'' = f(x,u,v), \qquad x \in (a,b)$$
$$v^{(4)} = g(x,u,v), \qquad x \in (a,b).$$

It would be interesting to derive Lyapunov-type inequalities for this kind of system.

5.1.5 Other Improvements

Let us recall that Patula proved in Lemma 2.1 for a single equation $-u'' = w(x)u$, with $u(a) = u(b) = 0$, the inequalities

$$\int_a^c w^+(x)dx > \frac{1}{c-a}, \qquad \int_c^b w^+(x)dx > \frac{1}{b-c},$$

which gives

$$\int_a^b w^+(x)dx > \frac{b-a}{(b-c)(c-a)}.$$

Lyapunov's inequality follows now by the arithmetic–geometric inequality.

There are many papers devoted to extensions of Theorem 5.1 considering system (1.6) following these lines, where inequalities like

$$\prod_{i=1}^{k}[(c_i - a)^{1-p_i} + (b - c_i)^{1-p_i}]^{\frac{\alpha_i}{p_i}} \leq \prod_{i=1}^{k}\left(\int_a^b w_i(x)dx\right)^{\frac{\alpha_i}{p_i}}$$

are proved, where $|u(c_i)| = \max_{x \in (a,b)} |u(c)|$, instead of the cleaner inequality

$$\frac{2^\alpha}{(b-a)^{\alpha-1}} \leq \prod_{i=1}^{k}\int_a^b w_i(x)dx,$$

which appears usually as a corollary.

Indeed, this is a minor improvement, since there is no information on the location of the points c_i.

However, for a single equation, there are some applications of this fact, and even ignoring the location of the maximum, it is possible to derive some interesting results in other contexts.

For example, let us introduce the so-called Fučík eigenvalue problem,

$$-u'' = \lambda u^+ - \mu u^- \qquad x \in (a,b),$$

with Dirichlet boundary conditions $u(a) = u(b) = 0$. The spectrum Σ is given by pairs (λ, μ), and in a few cases it has been proved that a sequence of continuous curves of eigenvalues $\lambda = f(\mu)$ belongs to Σ. Moreover, the first nontrivial curve, where the eigenfunction u changes sign once, approaches the trivial curves of the spectrum, the lines (λ_1, μ) or (λ, λ_1) when $\mu \to \infty$ or $\lambda \to \infty$, with λ_1 being the first Dirichlet eigenvalue of $-u'' = \lambda u$ and Dirichlet boundary conditions $u(a) = u(b) = 0$.

The picture for the Neumann boundary condition is different, since the trivial part of the spectrum is given by the axis $(\lambda, 0)$ and $(0, \mu)$, and the other Fučík eigenvalues remains separated from the trivial spectrum, a result proved first in [34] and obtained with this kind of technique in [95].

There are few works devoted to the Fučík spectrum for systems, and this is an area where similar inequalities, with Neumann boundary conditions instead of Dirichlet, seems to be applicable.

5.1.6 A Word of Caution

A quick search with the words *Lyapunov, inequality, systems* returns a large number of papers, and it is almost impossible to review all of them. There are some unsound claims of originality, and several improvements seems to be dubious. Worse, several papers are simply wrong. Of course, there are minor errors in almost every work (certainly including this one), such as a missing power in a formula or a multiplicative constant that has inadvertently been changed from one line to the next, but we are talking here about fatal mistakes. We mention here one of them that can be found in the literature.

5.1.6.1 A Frequent Mistake

The following mistake can be found in the literature on Lyapunov's inequalities, both for single equations and systems. The authors set out from

$$|u(t)|^p \le (t-a)^{p-1} \int_a^t |u'(x)|^p dx,$$

$$|u(t)|^p \le (b-t)^{p-1} \int_t^b |u'(x)|^p dx,$$

as in Remark 2.7, and now, since

$$t-a \le 2\frac{(b-t)(t-a)}{b-a} \qquad \text{when } a \le t \le \frac{a+b}{2},$$

$$b-t \le 2\frac{(b-t)(t-a)}{b-a} \qquad \text{when } \frac{a+b}{2} \le t \le b,$$

they obtain

$$|u(t)|^p \le \left(2\frac{(b-t)(t-a)}{b-a}\right)^{p-1} \int_a^{\frac{a+b}{2}} |u'(x)|^p dx,$$

$$|u(t)|^p \le \left(2\frac{(b-t)(t-a)}{b-a}\right)^{p-1} \int_{\frac{a+b}{2}}^b |u'(x)|^p dx.$$

However, *it is incorrect to add both equations and say that*

$$2|u(t)|^p \le \left(2\frac{(b-t)(t-a)}{b-a}\right)^{p-1} \int_a^b |u'(x)|^p dx,$$

because different things have been given the same name.

Starting from this equation, several authors have obtained different versions of the inequality

$$\frac{(b-a)^{p-1}}{2^{p-2}} \le \int_a^b [(x-a)(b-x)]^{p-1} w(x) dx \tag{1.15}$$

that are false, as one can check by considering a family of weights w concentrated at the center of the interval, since $\delta_{(a+b)/2}(x)$ is the critical weight where the inequality is sharp: we get

$$\frac{(b-a)^{p-1}}{2^{p-2}} \le \left(\frac{b-a}{2}\right)^{2(p-1)},$$

or equivalently, $2^p \le (b-a)^{p-1}$. Clearly, this is not true for the interval $[0,1]$, and let us note that no restrictions have been imposed on the length of the interval.

Although inequality (1.15) seems to be the nonlinear analogue of the Hartman and Wintner inequality in [63], the correct version for p-Laplacian problems can be deduced from Tang and He [108] in the previous subsection, and we have

$$1 \le \int_a^b w(t) \frac{[(b-t)(t-a)]^{p-1}}{(t-a)^{p-1}+(b-t)^{p-1}} dt.$$

5.2 Nehari–Calogero–Cohn and Resonant Systems

We consider now an extension of the Nehari–Calogero–Cohn inequality for resonant systems. We will consider the system

$$\begin{cases} -(|u'|^{p-2}u')' = \lambda w(x)\alpha|u|^{\alpha-2}u|v|^{\beta} & x \in (a,b), \\ -(|v'|^{q-2}v')' = \lambda w(x)\beta|u|^{\alpha}|v|^{\beta-2}v & x \in (a,b), \\ u(a) = u(b) = v(a) = v(b) = 0, \end{cases} \quad (2.16)$$

and we assume that

$$\frac{\alpha}{p} + \frac{\beta}{q} = 1, \quad \text{and} \quad 1 < q \le p < \infty. \quad (2.17)$$

We have the following result:

Theorem 5.3. *Let λ_1 be the first eigenvalue of problem (2.16), with $w \in L^1(a,b)$ a nonnegative monotonic function, and let us suppose that condition (2.17) holds. Then*

$$\left(\alpha^{\frac{\alpha}{p}}\beta^{\frac{\beta}{q}}\right)^{\frac{1}{\alpha+\beta}} \le \lambda^{\frac{1}{\alpha+\beta}} \int_a^b w^{\frac{1}{\alpha+\beta}}(x)dx.$$

Proof. Using the Rayleigh quotient associated to λ_1,

$$\lambda_1 = \sup_{(u,v)\in W_0} \frac{\int_a^b u^{\alpha}(x)v^{\beta}dx}{\int_a^b u'^p(x)+v'^q dx},$$

where

$$W_0 = \{(u,v) \in W_0^{1,p}(a,t_1) \oplus W_0^{1,q}(a,t_1)\},$$

we can repeat the arguments of the scalar case, and it is enough to compute the first eigenvalue of the same system when the weight is given by

$$s(x) = \begin{cases} \sigma^{\alpha+\beta} & \text{if } x \in [a,t_1], \\ 0 & \text{if } x \in (t_1,b), \end{cases}$$

for some $t_1 \in (a,b)$, $\sigma > 0$, with mixed boundary conditions

$$u(a) = v(a) = 0, \quad u'(t_1) = v'(t_1) = 0.$$

That is,

$$\Lambda_1 = \sup_{(u,v)\in W} \frac{\int_a^{t_1} u^{\alpha}(x)v^{\beta}dx}{\int_a^{t_1} u'^p(x)+v'^q dx},$$

where
$$W = \{(u,v) \in W^{1,p}(a,t_1) \oplus W^{1,q}(a,t_1) \; : \; u(a) = v(a) = 0\}.$$

However, we are unable to compute this eigenvalue, so we will use a lower bound for it. Let us recall first that Lyapunov's inequality gives, when $w \equiv 1$,

$$1 \le \alpha^{\frac{\alpha}{p}} \beta^{\frac{\beta}{q}} (t_1 - a)^{\frac{\alpha}{p'} + \frac{\beta}{q'}} (t_1 - a)^{\frac{\alpha}{p} + \frac{\beta}{q}} \Lambda_1. \tag{2.18}$$

Thus we have

$$\lambda_1^{-1} \sigma^{-(\alpha+\beta)} \le \frac{\int_a^{t_1} u^\alpha(x) v^\beta \, dx}{\int_a^{t_1} u'^p(x) + v'^q dx} \le \Lambda_1 \le \alpha^{\frac{\alpha}{p}} \beta^{\frac{\beta}{q}} (t_1 - a)^{\frac{\alpha}{p'} + \frac{\beta}{q'} + 1}.$$

Let us observe that

$$\frac{\alpha}{p'} + \frac{\beta}{q'} + 1 = \frac{\alpha(p-1)}{p} + \frac{\beta(q-1)}{q} = \alpha - \frac{\alpha}{p} + \beta - \frac{\beta}{q} + 1 = \alpha + \beta,$$

and therefore,

$$\left(\alpha^{\frac{\alpha}{p}} \beta^{\frac{\beta}{q}}\right)^{\frac{1}{\alpha+\beta}} \le \lambda^{\frac{1}{\alpha+\beta}} \sigma(t_1 - a) = \lambda^{\frac{1}{\alpha+\beta}} \int_a^b s^{\frac{1}{\alpha+\beta}}(x) dx.$$

The proof is finished. □

Remark 5.1. Although the trick we have used involves Lyapunov's inequality, we have applied it only to a constant-coefficient problem, so the weight w might not belong to L^1, provided that $w^{\frac{1}{\alpha+\beta}} \in L^1(a,b)$.

Remark 5.2. We believe that the constant in this lower bound can be improved.

Remark 5.3. This bound suggests a striking result: when the interval (a,b) collapses, then after rescaling, the system converges to a single equation, which is a $\alpha + \beta$-Laplacian. The details will appear elsewhere, since they are beyond the scope of this book.

5.3 Lyapunov-Type Inequalities in R^N

The subject of this section is a recent extension of the Lyapunov inequality to quasi-linear operators with Dirichlet boundary conditions defined on N-dimensional domains; see [37].

Let us remark that Egorov and Kondratiev obtained some inequalities for linear higher-order operators with Dirichlet boundary conditions [42], although the authors never related their inequalities to the Lyapunov inequality. Moreover, they derived some bounds for linear second-order operators with a different power of the weight. Such bounds were extended recently to Neumann boundary conditions by Cañada, Montero, and Villegas in [19, 22].

On the other hand, for quasilinear problems, Anane [2] and Cuesta [32] obtained some lower bounds for the first eigenvalue of p-Laplacian operators, involving some norm of the weight and the measure of the domain, which can be thought of as Lyapunov-type inequalities.

We will review briefly those inequalities below, except the work of Cañada et al., which was surveyed recently in [21].

The inequalities introduced in [37] were initially motivated by the following observation: in the classical Lyapunov inequality (1.2) and its extension to p-Laplacian problems (2.9),

$$\frac{4}{b-a} \leq \int_a^b w(x)\,dx, \quad \frac{2^p}{(b-a)^{p-1}} \leq \int_a^b w(x)\,dx,$$

we can interpret $(b-a)$ as the *length* of the interval; however, we can rewrite it as

$$\frac{2}{(b-a)/2} \leq \int_a^b w(x)\,dx, \quad \frac{2}{[(b-a)/2]^{p-1}} \leq \int_a^b w(x)\,dx,$$

and now we may think of $(b-a)/2$ as the *inner radius* of the interval.

Let us recall that the inner radius r_Ω of a set $\Omega \subset R^N$, $N \geq 1$, is defined as

$$r_\Omega = \sup_{x \in \Omega} \inf_{y \in \partial\Omega} |x-y|,$$

that is, the radius of the greatest open ball that we can place inside Ω.

Hence, the inequalities obtained in [37] for the problem

$$\begin{cases} -\operatorname{div}(|\nabla u|^{p-2}\nabla u) = \lambda w(x)|u|^{p-2}u, & x \in \Omega \\ u(x) = 0, & x \in \partial\Omega, \end{cases} \tag{3.19}$$

involve the inner radius of Ω instead of the measure of Ω as in the works of Anane and Cuesta, and for this reason, we obtain lower bounds for the first eigenvalue that improve in certain cases those obtained with other methods such as isoperimetric bounds and comparison using Sturmian arguments.

In the proofs, we will use as usual $s' = s/(s-1)$ and $p' = p/(p-1)$ to denote the Hölder conjugate exponents of s and p, and p^* will denote the critical exponent in the Sobolev immersion, $p^* = Np/(N-p)$ for $p < N$.

5.3.1 Related Inequalities

5.3.1.1 Egorov and Kondratiev Estimates

The following theorem can be found in [42] without proof.

Theorem 5.4. *Let* $\lambda(w)$ *be the first eigenvalue of the problem*

$$(-\Delta)^m u = \lambda w(x)u, \qquad u \in H_0^m(\Omega),$$

where $w \in L^\infty(\Omega)$ *is a positive weight,* $\Omega \subset R^N$ *is a bounded open set, and* $m \geq 1$.
Let us introduce, for $\alpha \in R$,

$$R_\alpha = \{w \in L^\infty(\Omega) : \int_\Omega w^\alpha(x)dx = 1\},$$

$$m_\alpha = \inf_{w \in R_\alpha}\{\lambda(w)\}.$$

Then:

(a) If $N > 2m$ *and* $\alpha > N/2m$, *we have* $m_\alpha > 0$.
(b) If $N > 2m$ *and* $\alpha \leq N/2m$, *we have* $m_\alpha = 0$.
(c) If $N < 2m$ *and* $\alpha \geq 1$, *we have* $m_\alpha > 0$.
(d) If $N < 2m$ *and* $\alpha < 1$, *we have* $m_\alpha = 0$.

We omit the proof of this theorem, which follows the same lines as those of the proofs of Anane and Cuesta below. They are based on the Sobolev immersion theorems and Morrey's lemma for Sobolev spaces $W_0^{2,m}$. Items (b) and (d) can be handled as in Sect. 5.3.3 by constructing appropriate weights.

5.3.1.2 Anane's and Cuesta's Estimates

We consider now the p-Laplacian eigenvalue problem (3.19). Anane obtained in [2] the following theorem:

Theorem 5.5. *Let* λ *be the first eigenvalue of problem (3.19), and assume that* $w \in L^\infty(\Omega)$. *Then there exists a fixed positive constant* C *such that*

$$\frac{C}{|\Omega|^\sigma \|w\|_\infty} \leq \lambda,$$

where

$$\sigma = p/N \qquad if \quad 1 < p \leq N,$$
$$\sigma = 1/2 \qquad if \quad N < p.$$

On the other hand, Cuesta proved in [32] the following result:

Theorem 5.6. *Let* λ *be the first eigenvalue of problem (3.19), and assume that* $w \in L^s(\Omega)$, *where*

$$s > N/p \qquad if \quad 1 < p \leq N,$$
$$s = 1 \qquad if \quad N < p.$$

Then there exists a fixed positive constant C *such that*

$$\frac{C}{|\Omega|^{\frac{sp-N}{sN}} \|w\|_s} \leq \lambda.$$

We condense both proofs into a single one, due to their similarity, although it is convenient to split the cases $p \leq N$ and $p > N$.

5.3.1.3 Case $p < N$

In Cuesta's proof, from the weak formulation of problem (3.19), with u as a test function, we have

$$\int_\Omega |\nabla u|^p \mathrm{d}x = \lambda \int_\Omega w(x)|u|^p \mathrm{d}x \leq \lambda \|w\|_s |\Omega|^{\frac{p^*-s'p}{s'p^*}} \left(\int_\Omega |u|^{p^*} \right)^{\frac{p}{p^*}},$$

where we have used Hölder's inequality with s, p^*/p, and $(s'p^*)/(p^*-s'p)$.

Now using Sobolev's inequality (A.1), we get

$$\int_\Omega |\nabla u|^p \mathrm{d}x = \lambda \int_\Omega w(x)|u|^p \mathrm{d}x \leq \lambda \|w\|_s |\Omega|^{\frac{p^*-s'p}{s'p^*}} \int_\Omega |\nabla u|^p \mathrm{d}x,$$

and the result follows.

Anane's proof is similar:

$$\int_\Omega |\nabla u|^p \mathrm{d}x = \lambda \int_\Omega w(x)|u|^p \mathrm{d}x \leq \lambda \|w\|_\infty |\Omega|^{1-\frac{p}{p^*}} \left(\int_\Omega |u|^{p^*} \right)^{\frac{p}{p^*}},$$

and using again Sobolev's inequality (A.1), the case $p < N$ is finished.

5.3.1.4 Case $p \geq N$

The proof when $p \geq N$ is different in both works. Anane proceeded as before,

$$\int_\Omega |\nabla u|^p \mathrm{d}x = \lambda \int_\Omega w(x)|u|^p \mathrm{d}x \leq \lambda \|w\|_\infty \int_\Omega |u|^p \mathrm{d}x,$$

and he used $q = 2p$ and Sobolev's inequality (A.1).

For $p = N$, Cuesta used instead the Sobolev immersion $W_0^{1,N} \subset L^{Ns'}$ in order to choose q, and using Hölder's inequality, she obtained $\|w\|_s$ instead of $\|w\|_\infty$.

The case $p > N$ follows from Morrey's Theorem A.2 and Poincaré's inequality:

$$\int_\Omega |\nabla u|^p \mathrm{d}x = \lambda \int_\Omega w(x)|u|^p \mathrm{d}x \leq \lambda \|u\|_\infty^p \int_\Omega w(x)\mathrm{d}x \leq \lambda \|w\|_1 \leq C \int_\Omega |\nabla u|^p \mathrm{d}x,$$

and the theorems are proved.

Remark 5.4. Let us observe that the constant C in the last inequality involves Poincaré's inequality, and hence it depends on the first eigenvalue of the p-Laplacian in Ω without weights. So we cannot obtain from them an explicit lower bound for the first eigenvalue.

5.3.2 Lyapunov-Type Inequalities Involving the Inner Radius

In [37] we proved the following result:

Theorem 5.7. *Let λ_1 be the first eigenvalue of*

$$-\Delta_p u = \lambda w(x)|u|^{p-2}u$$

in Ω with zero Dirichlet boundary conditions in $\partial\Omega$. Then

- *for $p > N$ and $w \in L^1(\Omega)$ with $w^+ \not\equiv 0$, we have*

$$\frac{C_1}{r_\Omega^{p-N}\|w^+\|_1} \leq \lambda_1;$$

- *for $p < N$ and $w \in L^s(\Omega)$ with $s > N/p$, $w^+ \not\equiv 0$, we have*

$$\frac{C_2}{r_\Omega^{\frac{sp-N}{s}}\|w^+\|_s} \leq \lambda_1.$$

The constant C_1 depends only on p and N, and the constant C_2 depends on p, N, and the constant C_h in Hardy's inequality.

Proof. We divide the proof into two steps.

1. Case $p > N$.
Let $u \in W_0^{1,p}(\Omega)$ be the first eigenfunction associated to λ_1. We have

$$\int_\Omega |\nabla u|^p = \lambda_1 \int_\Omega w(x)|u|^p.$$

Since $p > N$, Morrey's lemma (see Theorem A.2 in the appendix) implies that u is a continuous function, and $|u(x)|$ attains its maximum at some point $c \in \overline{\Omega}$.
Let us choose $x \in \partial\Omega$ such that

$$|x - c| = \inf_{x \in \partial\Omega} |x - c|,$$

and replacing $y = c$ in Morrey's inequality (see Eq. (2.3) in the appendix), we get

$$
\begin{aligned}
|u(c)| &\leq C(n,p)|x-c|^\alpha \left(\int_\Omega |\nabla u|^p \mathrm{d}x\right)^{\frac{1}{p}} \\
&= C(n,p)|x-c|^\alpha \left(\lambda_1 \int_\Omega w(x)|u|^p \mathrm{d}x\right)^{\frac{1}{p}} \\
&\leq C(n,p)|x-c|^\alpha \left(\lambda_1 \int_\Omega w^+(x)\mathrm{d}x\right)^{\frac{1}{p}} |u(c)|.
\end{aligned}
$$

We use now that $|x - y| \leq r_\Omega$, the inner radius of Ω, and by raising both sides to the power p after canceling $|u(c)|$, we obtain

$$1 \leq r_\Omega^{p-N} \lambda_1 C(N,p)^p \int_\Omega w^+(x)dx.$$

Since $p\alpha = p - N$, we get

$$\frac{C(N,p)^{-p}}{r_\Omega^{p-N} \int_\Omega w^+(x)dx} \leq \lambda_1,$$

and the proof for $p > N$ is finished.

2. Case $p < N$.

$$q = \alpha p + (1 - \alpha)p^*,$$

The proof begins now with the following simple inequality:

$$\frac{1}{r_\Omega^{\alpha p}} \int_\Omega |u|^{\alpha p} |u|^{(1-\alpha)p^*} dx \leq \int_\Omega \frac{|u|^{\alpha p + (1-\alpha)p^*}}{d(x)^{\alpha p}} dx,$$

where $\alpha \in (0,1)$ will be determined later. From Hölder's inequality with exponents $1/\alpha$ and $(1/\alpha)' = 1/(1-\alpha)$, we obtain

$$\frac{1}{r_\Omega^{\alpha p}} \int_\Omega |u|^{\alpha p + (1-\alpha)p^*} dx \leq \left(\int_\Omega \frac{|u|^p}{d(x)^p} dx \right)^\alpha \left(\int_\Omega |u|^{p^*} dx \right)^{1-\alpha},$$

and by the Hardy and Sobolev inequalities (see inequalities (2.4) and (2.3) in the appendix), we get

$$\frac{1}{r_\Omega^{\alpha p}} \int_\Omega |u|^{\alpha p + (1-\alpha)p^*} dx \leq C_{hs} \left(\int_\Omega |\nabla u|^p dx \right)^{\frac{\alpha + (1-\alpha)p^*}{p}},$$

where the constant C_{hs} depends on the constants C_h and C_s involved in the Hardy and Sobolev inequalities.

Since

$$\int_\Omega |\nabla u|^p = \lambda_1 \int_\Omega w(x)|u|^p,$$

we substitute into the previous inequality, and Hölder's inequality gives

$$\frac{1}{r_\Omega^{\alpha p}} \int_\Omega |u|^{\alpha p + (1-\alpha)p^*} dx \leq C_{hs} \left(\int_\Omega |u|^{ps'} dx \right)^{\frac{\alpha p + (1-\alpha)p^*}{ps'}}$$

$$\cdot \left(\int_\Omega \lambda_1^s w(x)^s dx \right)^{\frac{\alpha p + (1-\alpha)p^*}{ps}}.$$

We choose now α such that $ps' = \alpha p + (1 - \alpha)p^*$. Hence

$$\frac{\alpha p + (1 - \alpha)p^*}{ps} = \frac{s'}{s}$$

and

$$\alpha = \frac{p^* - ps'}{p^* - p},$$

so we obtain

$$\frac{1}{r_{\Omega}^{\alpha p}} \int_{\Omega} |u|^{\alpha p + (1-\alpha)p^*} \, dx \leq \|\lambda_1 w\|_{L^s}^{s'} \int_{\Omega} |u|^{\alpha p + (1-\alpha)p^*} dx,$$

where the exponent of the inner radius is

$$\frac{\alpha p}{s'} = \frac{sp - N}{s}.$$

The proof is finished. $\square$

Remark 5.5. Let us note that Theorem 5.7 does not cover the case $p = N$. To our knowledge, such an extension is not possible in this case due to the following results of Osserman.

Theorem 5.8 (Osserman [90]). *Let $\Omega \in R^2$ be an open set with connectivity $k \geq 2$. Then the first Dirichlet eigenvalue of the problem*

$$\begin{cases} -\Delta u = \lambda u \ in \ \Omega, \\ \quad u = 0 \quad on \ \partial\Omega, \end{cases}$$

satisfies

$$\lambda_1 \geq \frac{1}{k^2 r_{\Omega}^2}.$$

Theorem 5.9 (Osserman [90]). *Let $\Omega \in R^2$, and let $\Omega_\varepsilon = \Omega \setminus \cup_{i=1}^{j} B(x_i, \varepsilon)$, where $\{x_i\}_{i=1}^{j} \subset \Omega$. Then*

$$\lim_{\varepsilon \to 0} \lambda_1(\Omega_\varepsilon) = \lambda_1(\Omega).$$

Clearly, the inner radius of Ω can be made arbitrarily small by placing many points $\{x_i\}_{i=1}^{j}$ close enough.

However, we can expect some kind of inequalities in this case. Let us also mention that the extension of these theorems to quasilinear problems has not yet been made.

5.3.3 Optimality of the Powers

We show now that the power in the previous inequalities cannot be improved. To this end, it is enough to work with balls $B \subset R^N$ with radius r.

Theorem 5.10. *Let $B(0,r)$ be the ball of radius r centered at the origin, and let us set*

$$\gamma = \begin{cases} p - N & \text{if } p > N, \\ \dfrac{sp - N}{s} & \text{if } p < N. \end{cases}$$

- *Let $r > 1$. For $\beta < \gamma$ and C fixed, there exist a nonnegative weight w and $u_\beta \in W_0^{1,p}(B(0,r))$ a nontrivial solution of*

$$\begin{cases} -\Delta_p u = w(x)|u|^{p-2}u & \text{in } B(0,r), \\ u = 0 & \text{on } \partial B(0,r), \end{cases}$$

such that the inequality

$$\frac{C}{r^\beta} \leq \|w\|_{L^1(B(0,r))}$$

does not hold.

- *Let $r < 1$. For $\beta > \gamma$ and C fixed, there exist a nonnegative weight w and a solution $u_\beta \in W_0^{1,p}(B(0,r))$ of*

$$\begin{cases} -\Delta_p u = w(x)|u|^{p-2}u & \text{in } B(0,r), \\ u = 0 & \text{on } \partial B(0,r), \end{cases}$$

such that the inequality

$$\frac{C}{r^\beta} \leq \|w\|_{L^1(B(0,r))}$$

does not hold.

The case $p > N$, $r > 1$ can be found in [37], and the remaining cases follow exactly in the same way. Let us sketch the proof for $p > N$, $r < 1$; we shall show that the bound cannot hold for any power $\beta > p - N$.

Let us take $\varepsilon < 1 < r$ and

$$w(\rho) = \chi_{[0,\varepsilon]}(\rho)\rho^{1-N},$$

where $\chi_{[0,\varepsilon]}(\rho) = 1$ if $\rho \in [0,\varepsilon]$, and zero outside. We have

$$\|w\|_1 = \int_{B(0,r)} \chi_{[0,\varepsilon]}(|x|)|x|^{1-N}dx$$

$$= \int_{\omega_{N-1}} \int_0^\varepsilon \rho^{1-N}\rho^{N-1}d\rho d\theta$$

$$= \omega_{N-1}\varepsilon,$$

where ω_{N-1} is the surface measure of the unit ball.

Let us denote by $\lambda_1^{(r)}$ and $\lambda_1^{(\varepsilon)}$ the first eigenvalues of the p-Laplacian problems

$$-\Delta_p u = \lambda w(x)|u|^{p-2}u$$

with Dirichlet boundary conditions in $B(0,r)$ and $B(0,\varepsilon)$ respectively.

The inclusion $W_0^{1,p}(B(0,\varepsilon)) \subset W_0^{1,p}(B(0,r))$, which follows by extending each function by zero outside $B(0,\varepsilon)$, implies that

$$\lambda_1^{(r)} < \lambda_1^{(\varepsilon)},$$

according to the variational characterizations

$$\lambda_1^{(r)} = \inf_{\{u \in W_0^{1,p}(B(0,r)): u \neq 0\}} \frac{\int_{B(0,r)} |\nabla u|^p dx}{\int_{B(0,r)} \chi_{[0,\varepsilon]}(|x|)|x|^{1-N}dx},$$

$$\lambda_1^{(\varepsilon)} = \inf_{\{u \in W_0^{1,p}(B(0,\varepsilon)): u \neq 0\}} \frac{\int_{B(0,\varepsilon)} |\nabla u|^p dx}{\int_{B(0,\varepsilon)} |x|^{1-N}dx}.$$

We use now that the first eigenfunction in a ball is radial, which gives

$$\lambda_1^{(r)} \leq \lambda_1^{(\varepsilon)}$$

$$= \inf_{\{u \in W^{1,p}(0,\varepsilon): u(\varepsilon)=0, u \neq 0\}} \frac{\int_0^\varepsilon \rho^{N-1}|u'|^p d\rho}{\int_0^\varepsilon |u|^p d\rho}$$

$$\leq \varepsilon^{N-1} \frac{\pi_p^p}{\varepsilon^p}.$$

So, if the inequality

$$\frac{C}{r^\beta} \leq \lambda_1 \omega_{N-1} \varepsilon,$$

holds, we must have

$$\frac{C}{r^\beta} \leq \pi_p^p \omega_{N-1} \varepsilon^{N-p}.$$

We take now $r = \varepsilon^{-\alpha}$ with $\alpha > 0$, and we get

$$\frac{C}{\pi_p^p \omega_{N-1}} \leq \varepsilon^{N-p+\alpha\beta}. \tag{3.20}$$

We choose α such that $N - p + \alpha\beta > 1$, that is,

$$\alpha > \frac{p-N}{\beta},$$

which is positive, since $p > N$, and greater than 1 since $\beta > p - N$. Clearly, for ε sufficiently small, we obtain a contradiction, since the left-hand side of inequality (3.20) is a fixed constant, and the right-hand side can be made arbitrarily small.

5.3.4 An Incomplete Landscape

Given an elliptic operator L of order $2m$ and homogeneity p with zero Dirichlet boundary conditions, we may ask about the interplay among the following:

- The order $2m$ of the elliptic operator L.
- The exponent p of the corresponding Sobolev space $W_0^{m,p}$.
- The spatial dimension N.
- The appropriate power s of w appearing in the inequality.

A partial answer is given by Theorems 5.4 and 5.7, and it is not difficult to imagine a general result covering all the cases.

Moreover, at least in the linear case, we can expect an extension to fractional operators, and also to differential operators defined on sets with noninteger dimension, or with measures as weights.

There are several problems along these lines that deserve to be studied.

5.4 Lyapunov-Type Inequalities and Orlicz Spaces

5.4.1 Preliminaries

In this section, we consider the following φ-Laplacian equation:

$$- (\varphi(u'))' = w(x)\varphi(u), \qquad x \in (a,b), \tag{4.21}$$

with zero Dirichlet boundary conditions $u(a) = u(b) = 0$. Here $\varphi : R \to R$ is an odd nondecreasing function, convex in $(0,\infty)$ and vanishing only at zero. The main results of this section were proved in [35], and we also include some results from [104].

We say that a real function u is a solution of Eq. (4.21) if u is of class C^1, and $\varphi(u')$ is absolutely continuous, satisfying Eq. (4.21) almost everywhere.

The functional setting for this kind of problem is the Orlicz–Sobolev spaces $W_0^1 L_\Phi(a,b)$, where

$$\Phi(t) = \int_0^{|t|} \varphi(s)\mathrm{d}s$$

is a Young or N function, that is, an even, convex, continuous, and positive function, satisfying

$$\lim_{t \to 0} \frac{\Phi(t)}{t} = 0, \qquad \lim_{t \to \infty} \frac{\Phi(t)}{t} = \infty.$$

The Orlicz class $\widehat{L_\Phi}(a,b)$ is defined as

$$\widehat{L_\Phi}(a,b) = \left\{ u \text{ measurable } : \int_a^b \Phi(u) dx < \infty \right\},$$

and $L_\Phi(a,b)$ is the convex hull of this class, which is a Banach space with the Luxemburg norm

$$\|u\|_\Phi = \inf \left\{ \alpha > 0 : \int_a^b \Phi\left(\frac{u}{\alpha}\right) dx \leq 1 \right\}.$$

We define the Orlicz–Sobolev space $W^1 L_\Phi(a,b)$ as

$$W^1 L_\Phi(a,b) = \{ u \in L_\Phi(a,b) : u' \in L_\Phi(a,b) \},$$

with the norm

$$\|u\|_{1,\Phi} = \|u\|_\Phi + \|u'\|_\Phi.$$

As usual, $W_0^1 L_\Phi(a,b)$ is the closure of $C_0^\infty(a,b)$ in $W^1 L_\Phi(a,b)$ with the norm $\|u\|_{1,\Phi}$.

We omit here the technical details concerning Sobolev–Orlicz immersions and dual spaces and the existence and uniqueness of solutions. We present only some facts about the eigenvalue problem; see [55, 56, 83, 109] and the references therein for details.

However, in this work we will need an additional hypothesis on $\psi(s) = s \cdot \varphi(s)$, which is also convex, the so-called Δ_2 condition:

Definition 5.1. We say that a Young function Φ satisfies the Δ_2 condition if there exist some $k > 0$ and x_0 such that

$$\Phi(2x) \leq k\Phi(x)$$

for every $x \geq x_0$. If we can take $x_0 = 0$, we will say that Φ satisfies a global Δ_2 condition.

Remark 5.6. Let us note that if $\Phi(x)$ satisfies the Δ_2 condition, we have

$$\Phi(x) \leq x\varphi(x) \leq \Phi(2x),$$

and also $\psi(s) = s \cdot \varphi(s)$ satisfies the Δ_2 condition.

Let us denote by $\bar\Phi$ the conjugate function of Φ, which is defined by

$$\bar\psi(t) = \sup_{s \in R} \{ ts - \Phi(s) \}.$$

Different approaches to the eigenvalue problem for φ-Laplacians can be found in [55, 56, 109]. When both Φ and $\bar{\Phi}$ satisfy the Δ_2 condition, $W_0^1 L_\Phi(a,b)$ is a separable and reflexive Banach space, and we can apply the Lyusternik–Schnirelmann theory sketched in Sect. A.3.3.2 in the appendix. So for fixed $r > 0$, we obtain a sequence of eigenvalues $\{\lambda_k(r)\}_{k \geq 1}$, with corresponding eigenfunctions $\{u_k\}_{k \geq 1} \subset W_0^1 L_\Phi(a,b)$, such that

$$-(\varphi(u_k'))' = \lambda_k(r)\varphi(u_k),$$

$$\lambda_k(r) = \inf_{C \in \mathscr{C}_k} \sup_{u \in C} \int_a^b \Phi(u')dx,$$

where

$$C = \left\{ u \in W_0^1 L_\Phi(a,b) \; : \; \int_a^b \Phi(u)dx = r \right\},$$

and $\mathscr{C}_k$ is the class of sets of genus greater than or equal to k.

Without the Δ_2 condition, the problem is subtler and we need to use different techniques (subdifferentials as in [56], Galerkin approximations as in [109]) to obtain a sequence of eigenvalues.

An interesting point here is the dependence of those variational eigenvalues on the normalization of the Orlicz norm of u, i.e., the eigenfunctions satisfy the condition

$$\int_a^b \Phi(u_k)dx = r,$$

so we may ask whether

$$\lambda = \inf_{r>0}\{\lambda_1(r)\}$$

is bounded away from zero. A positive answer was given in [55] when Φ satisfies the Δ_2 condition or the domain is small. The proof given there follows from a Poincaré inequality for Orlicz–Sobolev spaces. Here, a Lyapunov-type inequality gives a quick answer to this question in the one-dimensional case and provides a computable lower bound for $\lambda_1(r)$. We believe that an extension to n-dimensional problems is possible, following the lines of the previous section.

5.4.2 Lyapunov's Inequality for φ-Laplacian Equations

From now on, we will take the existence of solutions or eigenvalues for granted, and we use the equation that every eigenpair must satisfy, namely

$$-(\varphi(u'))' = \lambda w(x)\varphi(u), \qquad x \in (a,b).$$

Therefore, after multiplying by u and integrating by parts, using that $u(a) = u(b) = 0$, we get

$$\int_a^b \psi(u')dx = \lambda \int_a^b \psi(u)dx. \tag{4.22}$$

The main result of this section is the following theorem:

Theorem 5.11. *Let $\psi(s) = s \cdot \varphi(s)$ be a convex nondecreasing function satisfying the Δ_2 condition with a positive constant k, for every $x \geq 0$. Let $w(x) \in L^1(a,b)$ be a positive integrable function, and let us suppose that there exists a nontrivial solution of Eq. (4.21) with zeros at a and b. Then*

$$C_{k,(b-a)} \leq \int_a^b w(x)dx, \tag{4.23}$$

where the constant is given by

$$C_{k,(b-a)} = 2 \left(\frac{k}{2}\right)^{[1-\log_2(b-a)]},$$

where $[x]$ is the greatest integer less than or equal to x.

The proof is based on the following simple lemma:

Lemma 5.1. *Let $\psi(s) = s \cdot \varphi(s)$ be an increasing convex function satisfying the Δ_2 condition for a fixed $k > 0$ and every $s \in R$. Then for $n \in Z$ and $2^n \leq c < 2^{n+1}$,*

$$c \left(\frac{k}{2}\right)^n \leq \frac{\psi(ct)}{\psi(t)}.$$

Moreover, the constant k is greater than or equal to 2.

Proof. Let us prove only the case $n \geq 0$, the other one being similar.

We observe first that since φ is nondecreasing and $\psi(s) = s \cdot \varphi(s)$, we have for $x \leq y$,

$$\frac{\psi(x)}{x} = \varphi(x) \leq \varphi(y) = \frac{\psi(y)}{y}.$$

Let us suppose first that $1 \leq c < 2$. Then

$$\frac{\psi(cx)}{cx} \geq \frac{\psi(x)}{x},$$

which implies

$$\frac{\psi(ct)}{\psi(t)} \geq c.$$

We assume now that the result holds for $c < 2^n$, and we take $c \in [2^n, 2^{n+1})$. We have

$$\frac{\psi(ct)}{\psi(t)} \geq \frac{k\psi(ct)}{\psi(2t)} = \frac{k\psi(2tc/2)}{\psi(2t)} \geq k \cdot \frac{c}{2} \left(\frac{k}{2}\right)^{n-1},$$

where the last inequality follows from the induction hypotheses.

Finally, since

$$\frac{\psi(x)}{x} \leq \frac{\psi(2x)}{2x} \leq k\frac{\psi(x)}{2x},$$

we get that $k \geq 2$, and the lemma is proved. $\square$

We are ready to prove Theorem 5.11.

Proof (Proof of Theorem 5.11). Let us take $c \in [a,b]$, where $|u|$ is maximized, and we can write

$$2|u(c)| = \left|\int_a^c u'(x)dx\right| + \left|\int_c^b u'(x)dx\right| \leq \int_a^b |u'(x)|dx.$$

We need to divide by $b - a$ in order to apply Jensen's inequality, and we get

$$\psi\left(\frac{2u(c)}{b-a}\right) = \psi\left(\frac{1}{b-a}\int_a^b |u'|dx\right) \leq \frac{1}{b-a}\int_a^b \psi(u')dx.$$

Now from Eq. (4.22), we obtain

$$\int_a^b \psi(u')dt = \int_a^b q(t)\psi(u)dt,$$

and substituting this in the previous inequality, we have

$$\psi\left(\frac{2u(c)}{b-a}\right) \leq \frac{1}{b-a}\int_a^b \psi(u')dx$$

$$= \frac{1}{b-a}\int_a^b w(x)\psi(u)dx$$

$$\leq \frac{\psi(u(c))}{b-a}\int_a^b w(x)dx,$$

where the last inequality follows because $|u(t)|$ reaches its maximum at c, and ψ is an even function.

We now apply Lemma 5.1 to obtain

$$\left(\frac{k}{2}\right)^n \leq \frac{\psi\left(\frac{2u(c)}{b-a}\right)}{\frac{\psi(u(c))}{b-a}} \leq \int_a^b w(x)dx, \tag{4.24}$$

with $n = [1 - \log_2(b-a)]$.

The proof is finished. $\square$

Remark 5.7. Theorem 5.11 gives the classical Lyapunov inequality for linear differential equations, with $\psi(s) = s^2$ and $k = 4$. Indeed, for the p-Laplacian, since $\psi(s) = s^p$ satisfies the hypotheses of the theorem with $k = 2^p$, we have that

$$C_{k,(b-a)} = 2\left(2^{p-1}\right)^{[1-\log_2(b-a)]}$$
$$= 2 \cdot 2^{([1-\log_2(b-a)](p-1))}$$
$$= 2 \cdot 2^{p-1} 2^{\log_2(b-a)^{-p+1}}$$
$$= \frac{2^p}{(b-a)^{p-1}}.$$

Remark 5.8. A large body of the literature on generalized Laplacians deals with asymptotically p-homogeneous functions with $p > 1$. That is, we say that $\psi(s)$ is asymptotically p-homogeneous if

$$\lim_{s \to +\infty} \frac{\psi(\sigma s)}{\psi(s)} = \sigma^{p-1}. \tag{4.25}$$

Here we can prove a Lyapunov-type inequality when $\|u\|_\infty$ is large enough, and the constant on the left-hand side approaches the constant of Lyapunov's inequality for the p-Laplacian. We have the following result.

Theorem 5.12. *Let* $\psi(s) = s \cdot \varphi(s)$ *be a convex, nondecreasing, asymptotically p-homogeneous function. Let us fix $\varepsilon > 0$. Then there exists m_ε such that*

$$\frac{2^p}{(b-a)^{p-1}} - \varepsilon \leq C_{k,(b-a)} \leq \frac{2^p}{(b-a)^{p-1}} + \varepsilon$$

whenever Eq. (4.21) admits a solution u with zeros at a and b, and $\|u\|_\infty \geq m_\varepsilon$.

Proof (Proof of Theorem 5.12). In the proof of Theorem 5.11 we obtained the following inequality:

$$\frac{\psi\left(\frac{2u(c)}{b-a}\right)}{\frac{\psi(u(c))}{b-a}} \leq \int_a^b w(x)\mathrm{d}x.$$

The condition (4.25) of asymptotic p-homogeneity implies that there exists m_ε such that

$$\sigma^{p-1} - \varepsilon \leq \frac{\varphi(\sigma s)}{\varphi(s)} \leq \sigma^{p-1} + \varepsilon$$

for $s \geq m_\varepsilon$, and therefore if $\|u\|_\infty \geq m_\varepsilon$, we get

$$\frac{2^p}{(b-a)^{p-1}} - \varepsilon \leq \frac{\psi\left(\frac{2u(c)}{b-a}\right)}{\frac{\psi(u(c))}{b-a}} = \frac{2\varphi\left(\frac{2u(c)}{b-a}\right)}{\varphi(u(c))} \leq \frac{2^p}{(b-a)^{p-1}} + \varepsilon.$$

The theorem is proved. □

5.4.3 Lower Bounds for Eigenvalues of φ-Laplacian Equations

As usual, we can replace w by $\lambda_k(r) \cdot w$ in Eq. (4.23), obtaining a lower bound for the kth eigenvalue, namely

$$\frac{C_{k,(b-a)}}{\int_a^b w(x)dx} \leq \lambda_k(r).$$

Moreover, Theorem 5.12 can be formulated in terms of the behavior of the eigen-curve $(\lambda_1(r)r)$ when $r \to \infty$. Let us note that if

$$r = \int_a^b \Phi(u)dx \to \infty,$$

then $\|u\|_\infty \to \infty$, and the spectrum is located (asymptotically) at the right-hand side of the line $x = 2^p(b-a)^{1-p} - \varepsilon$ for every $\varepsilon > 0$.

However, for solutions of eigenvalue problems with multiple nodal domains, we can improve the estimate:

Theorem 5.13. *Let $\psi(s)$ be a Young function satisfying the Δ_2 condition. Let u be an eigenfunction associated to some $\lambda(r)$ with m nodal domains. Then*

$$2m\left(\frac{k}{2}\right)^{[1+\log_2(m)-\log_2(b-a)]} \leq \lambda(r)\int_a^b w(x)dx.$$

Proof. Let us write the m nodal domains as $I_j = (x_{j-1}, x_j)$, for $1 \leq j \leq m$, and $x_0 = a$, $x_m = b$.

We can apply inequality (4.23) in each I_j (replacing w by $\lambda(r) \cdot w$), and we get

$$\lambda(r)\int_a^b w(x)dx = \lambda \sum_{j=1}^m \int_{x_{j-1}}^{x_j} w(x)dx$$

$$\geq \sum_{j=1}^m 2\left(\frac{k}{2}\right)^{[1-\log_2(x_j-x_{j-1})]}$$

$$= 2\frac{k}{2}\sum_{j=1}^m \left(\frac{k}{2}\right)^{-\log_2(x_j-x_{j-1})}$$

$$\geq 2m\left(\frac{k}{2}\right)^{1-\log_2(b-a)+\log_2(m)},$$

where in the last step we have used the arithmetic–geometric–harmonic mean inequality. The proof is finished. $\square$

Remark 5.9. In [35] there appeared an incorrect factor: $(b-a)$ instead of m.

Remark 5.10. When $\psi(s) = |s|^p$, we obtain the same bound as in Sect. 2.4 in Chap. 2, since

$$2m\left(2^{p-1}\right)^{1-\log_2(b-a)+\log_2(m)} = 2m2^{p-1}(b-a)^{-(p-1)}m^{(p-1)} = \frac{2^p m^p}{(b-a)^{p-1}}.$$

Remark 5.11. For asymptotically p-homogeneous functions φ, the existence of solutions with m nodal domains for a constant-coefficient φ-Laplacian was proved in [54]. As far as we know, for arbitrary functions φ, it is not known whether the kth eigenfunction has k nodal domains.

5.4.4 Some Generalizations

Let us note that the Δ_2 condition was needed to deal with the expression

$$\psi\left(\frac{2u(c)}{b-a}\right),$$

and we can relax the hypotheses of a global Δ_2 condition for solutions with $\|u\|_\infty$ big enough.

There is another hypothesis that is not completely equivalent to the Δ_2 condition (see [67] for an exhaustive analysis), the submultiplicative condition:

Definition 5.2. Let $J \subset [0,\infty)$ satisfy $J \cdot J \subset J$. We say that ψ is submultiplicative on $J \subset [0,\infty)$ if

$$\psi(x \cdot y) \leq \psi(x) \cdot \psi(y)$$

for every $x, y \in J$.

Remark 5.12. Clearly, if ψ is submultiplicative on J and $2 \in J$, then ψ satisfies the Δ_2 condition on J. However, there are functions ψ that satisfy the submultiplicative condition on $[0,1]$ but neither the submultiplicative condition on $[0,\infty)$ nor the Δ_2 condition.

With the aid of this condition, Sanchez and Vergara gave a simple proof avoiding the use of Jensen's inequality in [104]. We include it, since the hypotheses on φ are different, and a cleaner expression in terms of φ gives the lower bound:

Theorem 5.14. *Let $\varphi(s)$ be an odd, increasing, and submultiplicative function in $[0,\infty)$ such that $\phi(s) = 1/\varphi(s)$ is a convex function for every $s > 0$. Let w be a continuous function, and let u be a nontrivial solution of*

$$-(\varphi(u'))' = w(x)\varphi(u)$$

satisfying

$$u(a) = u(b) = 0,$$

and $u(x) \neq 0$ for every $x \in (a, b)$. Then

$$\frac{2}{\varphi\left(\frac{b-a}{2}\right)} \leq \int_a^b w(x)dx.$$

Proof. We can assume that $u > 0$ in (a, b) and that there exists $c \in (a, b)$ with $u'(c) = 0$. Since $\phi = 1/\varphi$ is convex, it follows that

$$\phi\left(\frac{b-a}{2}\right) = \frac{\phi(b-c)}{2} + \frac{\phi(c-a)}{2},$$

and we have

$$\frac{2}{\phi\left(\frac{b-a}{2}\right)} \leq \frac{1}{\phi(c-a)} + \frac{1}{\phi(b-c)}$$

$$= \frac{1}{\phi(u(c))}\left[\frac{\phi\left((c-a)\frac{u(c)}{c-a}\right)}{\phi(c-a)} + \frac{\phi\left((b-c)\frac{u(c)}{b-c}\right)}{\phi(b-c)}\right]. \tag{4.26}$$

Since ϕ is submultiplicative, we have

$$\frac{\phi\left((c-a)\frac{u(c)}{c-a}\right)}{\phi(c-a)} + \frac{\phi\left((b-c)\frac{u(c)}{b-c}\right)}{\phi(b-c)} \leq \phi\left(\frac{u(c)}{c-a}\right) + \phi\left(\frac{u(c)}{b-c}\right),$$

and the boundary condition $u(a) = u(b) = 0$ together with the fact that φ (and also ϕ) is an odd function enables us to write this last expression as

$$\phi\left(\frac{u(c) - u(a)}{c-a}\right) - \phi\left(\frac{u(b) - u(c)}{b-c}\right).$$

Since we are considering solutions $u \in C^1(a, b)$, the mean value theorem implies the existence of $x_1 \in (a, c)$ and $x_2 \in (c, b)$ such that

$$\frac{u(c) - u(a)}{c-a} = u'(x_1), \qquad \frac{u(b) - u(c)}{b-c} = u'(x_2),$$

and then

$$\phi\left(\frac{u(c) - u(a)}{c-a}\right) - \phi\left(\frac{u(b) - u(c)}{b-c}\right) = \phi(u'(x_1)) - \phi(u'(x_2))$$

$$= \int_c^{x_1} (\phi(u'(x)))'dx - \int_c^{x_2} (\phi(u'(x)))'dx$$

$$= -\int_{x_1}^c (\phi(u'(x)))'dx - \int_c^{x_2} (\phi(u'(x)))'dx$$

$$= -\int_{x_1}^{x_2} (\phi(u'(x)))' dx$$

$$\leq -\int_{a}^{b} (\phi(u'(x)))' dx$$

$$= \int_{a}^{b} w(x)\phi(u(x)) dx.$$

Returning to inequality (4.26), from the previous estimates we obtain

$$\frac{2}{\phi(\frac{b-a}{2})} \leq \frac{\int_{a}^{b} w(x)\phi(u(x)) dx}{\phi(u(c))}$$

$$\leq \int_{a}^{b} \frac{w(x)\phi(u(x))}{\phi(u(c))} dx$$

$$\leq \int_{a}^{b} \frac{w(x)\phi(u(x))}{\phi(u(x))} dx$$

$$= \int_{a}^{b} w(x) dx,$$

and the proof is finished. $\square$

Appendix A
Preliminaries

Abstract In this appendix we include some classical inequalities that will be needed in the book, and a short review of the basic theory of p-Laplace operators and some properties of their eigenvalues.

A.1 Some Basic Inequalities

We include here some inequalities for the sake of completeness. The proofs can be found in the classic book of Hardy et al. [60].

A.1.1 The Arithmetic–Geometric–Harmonic Mean Inequality

In several places we need the well-known arithmetic–geometric inequality

$$\sqrt{st} \leq \frac{s+t}{2}, \tag{1.1}$$

and we will use it mainly with $s = (b-c)$ and $t = (c-a)$.

This is only a small part of the arithmetic–geometric–harmonic mean inequality: given n positive numbers $\{a_n\}_{i=1}^n$, we have

$$n \cdot \left(\frac{1}{a_1} + \cdots + \frac{1}{a_n} \right)^{-1} \leq (a_1 \cdot \ldots \cdot a_n)^{\frac{1}{n}} \leq \frac{a_1 + \cdots + a_n}{n},$$

and equality holds only when all of the terms are equal, $a_1 = \cdots = a_n$.

Sometimes we will use the following variant of the arithmetic–geometric–harmonic mean inequality:

$$\frac{n^2}{a_1 + \cdots + a_n} \leq \frac{1}{a_1} + \cdots + \frac{1}{a_n}.$$

J.P. Pinasco, *Lyapunov-type Inequalities: With Applications to Eigenvalue Problems*, SpringerBriefs in Mathematics, DOI 10.1007/978-1-4614-8523-0,
© Juan Pablo Pinasco 2013

A.1.2 Minkowski's Inequality

There are different inequalities related to Minkowski. Let us mention here the following ones:

Minkowsi's inequality: given $f, g \in L^p(\Omega)$, $\Omega \subset R^N$, one has

$$\|f+g\|_p \leq \|f\|_p + \|g\|_p.$$

A companion to Minkowski's inequality, Theorem 27 in [60]: given n positive numbers $\{a_i\}_{i=1}^n$, we have

$$(a_1 + \cdots + a_n)^r > a_1^r + \cdots + a_n^r \qquad \text{for } r > 1,$$

$$(a_1 + \cdots + a_n)^r < a_1^r + \cdots + a_n^r \qquad \text{for } 0 < r < 1.$$

This inequality implies Theorem 199 in [60]: for positive functions f, g and $0 < r < 1$, we have

$$\int (f+g)^r \mathrm{d}x < \int f^r \mathrm{d}x + \int g^r \mathrm{d}x.$$

Minkowski's integral inequality, Theorem 202 in [60]: for $p \geq 1$,

$$\left(\int_A \left| \int_B F(x,t)\mathrm{d}t \right|^p \mathrm{d}x \right)^{\frac{1}{p}} \leq \int_B \left(\int_A |F(x,t)\mathrm{d}t|^p \mathrm{d}x \right)^{\frac{1}{p}} \mathrm{d}t.$$

A.1.3 A Useful Lemma

The arithmetic–geometric–harmonic mean inequality together with Minkowski's inequalities implies the following result:

Lemma A.1. *Given n positive numbers $\{a_i\}_{i=1}^n$, we have*

$$\min\left\{ \frac{1}{a_1^{p-1}} + \cdots + \frac{1}{a_n^{p-1}} \; : \; p \geq 1 \right\} = n \cdot \left(\frac{n}{\sum_{i=1}^n a_i} \right)^{p-1}.$$

A.1.4 Hölder's Inequality

Given $f \in L^p(\Omega)$, $g \in L^{\frac{p}{p-1}}(\Omega)$, $\Omega \subset R^N$, we have

$$\int_\Omega fg\mathrm{d}x \leq \left(\int_\Omega |f|^p\mathrm{d}x \right)^{\frac{1}{p}} \cdot \left(\int_\Omega |g|^{\frac{p}{p-1}} \mathrm{d}x \right)^{\frac{p-1}{p}}.$$

A.1.5 *Young's Inequality*

Given $f \in L^p(\Omega)$, $g \in L^{\frac{p}{p-1}}(\Omega)$, $\Omega \subset R^N$, we have

$$\int_\Omega fg dx \le \frac{1}{p}\int_\Omega |f|^p dx + \frac{p-1}{p}\int_\Omega |g|^{\frac{p}{p-1}} dx.$$

A.1.6 *Jensen's Inequality*

Given $f \in L^1(\Omega)$, $\Omega \subset R^N$, and $\varphi : R \to R$ a convex function, we have

$$\varphi\left(\frac{1}{|\Omega|}\int_\Omega f(x)dx\right) \le \frac{1}{|\Omega|}\int_\Omega \varphi(f(x))dx.$$

A.2 Sobolev Spaces and Related Inequalities

We recall the definition of the Sobolev spaces $W^{m,p}(\Omega)$ and $W_0^{m,p}(\Omega)$, for $1 \le p < \infty$ and $m \ge 1$. We refer the interested reader to [47] for details.

We say that $v \in L_{loc}^1(\Omega)$ is the kth weak partial derivative of $u \in L_{loc}^1(\Omega)$, $D^k u = v$, if

$$\int_\Omega u D^k \varphi dx = (-1)^k \int_\Omega v\varphi dx$$

for all test functions $\varphi \in C_0^\infty(\Omega)$, the space of C^∞ functions with compact support in Ω.

The Sobolev space $W^{m,p}(\Omega)$ is the set of functions $u \in L_{loc}^1(\Omega)$ such that for each multi-index $k = (k_1, \ldots, k_N)$ with $0 \le |k| \le m$, $D^k u$ exists and belongs to $L^p(\Omega)$.

The norm of $u \in W^{m,p}(\Omega)$ for $1 \le p < \infty$ is

$$\|u\|_{W^{m,p}(\Omega)} = \sum_{0 \le |k| \le m} \int_\Omega |D^k u|^p dx.$$

We denote by $W_0^{m,p}(\Omega)$ the closure of $C_0^\infty(\Omega)$ in $W_0^{m,p}(\Omega)$.

We will need the following classical inequalities for functions in Sobolev spaces.

Theorem A.1 (Sobolev's inequality). *Let* $\Omega \subset R^N$ *be a bounded open set, and* $u \in W_0^{1,p}$ *with* $1 \le p < N$. *Then*

$$\int_\Omega |u|^q dx \le C_{q,p}\left(\int_\Omega |\nabla u|^p\right)^{\frac{q}{p}} \tag{2.2}$$

with $1 \le q \le p^* = pN/(N-p)$.

Theorem A.2 (Morrey's lemma). *If $p > n$, there exists a constant C such that*

$$\|u\|_{C^{0,\alpha}(R^N)} \leq C \|u\|_{W^{1,p}(R^N)}.$$

In particular, there exists a constant $C(N,p)$ such that for all $u \in W_0^{1,p}(\Omega)$,

$$|u(x) - u(y)| \leq C(n,p)|x-y|^\alpha \|\nabla u\|_{L^p} \tag{2.3}$$

for all $x, y \in \overline{\Omega}$ and $\alpha = 1 - \frac{N}{p}$.

The proofs can be found, for example, in [47].

Theorem A.3 (Hardy's inequality). *Let $1 < p < N$, $u \in C_0^\infty(\Omega)$, and suppose that there exists a constant $\gamma > 0$ such that*

$$C_p((R^N \setminus \Omega) \cap \bar{B}(x,r), B(x,2r)) \geq \gamma C_p(\bar{B}(x,r), B(x,2r))$$

for every $x \in (R^N \setminus \Omega)$ and $r > 0$, where $C_p(K, B(x,2r))$ is the variational p-capacity

$$C_p(K,U) = \inf\left\{\int_U |\nabla u|^p dx : u \in C_0^\infty(U), u(x) \geq 1 \text{ for } x \in K\right\}.$$

Then there exists a constant C_h such that

$$\int_\Omega \frac{|u|^p}{d(x,\partial\Omega)^p} dx \leq C_h \int_\Omega |\nabla u|^p dx, \tag{2.4}$$

where $d(x,\partial\Omega)$ is the distance from $x \in \Omega$ to the boundary,

$$d(x,\partial\Omega) = \inf_{y\in\partial\Omega} |x-y|.$$

This theorem is due to Lewis [77]. Let us note that $C_{q,s}$ in Sobolev's inequality is a universal constant depending only on p, q, and N. However, C_h depends on the p-capacity of $R^N \setminus \Omega$, although for convex domains, we have $C_h = \left(\frac{p}{N-p}\right)^p$.

A.3 The p-Laplace Operator

In this section we review briefly the main results for the eigenvalue problem associated to the p-Laplace operator

$$-\Delta_p u := -\text{div}(|\nabla u|^{p-2}\nabla u),$$

where $1 < p < \infty$.

This is a quasilinear operator (also called a half-linear operator due to its homogeneity), which is singular if $p < 2$ and degenerate if $p > 2$. The eigenvalue problem is

$$-\mathrm{div}(|\nabla u|^{p-2}\nabla u) = \lambda w(x)|u|^{p-2}u, \qquad x \in \Omega \subset R^N,$$

and it was considered first by Browder [11–13], who proved the existence of a sequence of eigenvalues when $p \geq 2$. Later, several authors studied the eigenvalue problem, among them Amman and Fučík, (see [1, 52]), and then there was an explosion of research in the 1980s by Anane, de Thelin, Elbert, Lieb, Otani, Velin, to name only a few. For a more detailed survey, see [50]

A.3.1 The One-Dimensional Eigenvalue Problem

Let us consider the following eigenvalue problem for the p-Laplacian equation on (a,b):

$$- (|u'|^{p-2}u')' = \lambda w(x)|u|^{p-2}u, \tag{3.5}$$

with zero Dirichlet boundary conditions

$$u(a) = u(b) = 0. \tag{3.6}$$

Here, the weight $w(x) \in L^1(a,b)$ is a positive function, and $\lambda \in R$ is the eigenvalue parameter.

Although this problem can be analyzed in different ways (see [39]), we will use the fact that it has a variational structure, and Eq. (3.5) is the Euler–Lagrange equation of the functional

$$\phi(u) = \frac{1}{p}\int_a^b |u|^p \mathrm{d}x + \frac{1}{p}\int_a^b \lambda w(x)|u|^p \mathrm{d}x. \tag{3.7}$$

We will say that $\lambda \in R$ is an eigenvalue if there exists a nontrivial weak solution of problem (3.5), where by a weak solution we mean a critical point of the functional Eq. (3.7), a function $u \in W_0^{1,p}(a,b)$, and $\lambda \in R$ such that

$$\int_a^b |u'|^{p-2}u' \cdot v' \mathrm{d}x = \lambda \int_s^b w(x)|u|^{p-2}uv$$

for every test function $v \in W_0^{1,p}(a,b)$.

A.3.2 Constant-Coefficient Case

When w is a constant function, the eigenvalues and eigenfunctions can be computed explicitly (see [38, 57]). Let us denote by $\sin_p(x)$ the solution of the initial value problem

$$-(|u'|^{p-2}u')' = (p-1)|u|^{p-2}u,$$

$$u(0) = 0, \qquad u'(0) = 1,$$

and by direct integration, we can check that $\sin_p(\cdot)$ is defined implicitly as

$$x = \int_0^{\sin_p(x)} \frac{dt}{\sqrt[p]{1-t^p}} = 2\frac{\pi/p}{\sin(\pi/p)}.$$

Clearly, when $p = 2$, this is the definition of $\arcsin(\cdot)$.

Let us denote by $\hat{\pi}_p$ the first zero of $\sin_p(x)$, given by

$$\hat{\pi}_p = 2\int_0^1 \frac{dt}{\sqrt[p]{1-t^p}}.$$

Remark A.1. Let us observe that there are alternative definitions of $\sin_p$ and π_p, depending on the presence of the factor $p-1$ in the equation; see [41], for example. It is convenient to introduce

$$\pi_p = \sqrt[p]{p-1}\hat{\pi}_p$$

to recover an expression similar to the one that holds for linear problems.

With these definitions, $\sin_p(x)$ and $\sin_p'(x)$ satisfy

$$|\sin_p(x)| \leq 1, \quad |\sin_p'(x)| \leq 1.$$

Moreover, they satisfy a Pythagorean-like identity

$$|\sin_p(x)|^p + |\sin_p'(x)|^p = 1. \tag{3.8}$$

We have the following characterization of the spectrum:

Theorem A.4 (Del Pino et al. [38]). *The Dirichlet eigenvalues $\lambda_k(p)$ and eigenfunctions $u_{p,k}$ of problems (3.5)–(3.6) are given by*

$$\lambda_k(p) = \left(\frac{\pi_p k}{b-a}\right)^p, \qquad u_{p,k}(x) = \sin_p\left(\frac{\hat{\pi}_p kx}{b-a}\right).$$

Let us observe that the kth eigenvalue is simple, and the associated eigenfunction $u_{p,k}$ has k nodal domains, that is, $u_{p,k}$ has $k+1$ simple zeros in $[a,b]$.

Remark A.2. Similar computations show that the Neumann eigenvalues $\{\mu_k(p)\}_{k\geq 0}$ corresponding to the boundary condition $u'(a) = u'(b) = 0$ and the eigenvalues

$\{v_k(p)\}_{k \geq 1}$ corresponding to the mixed boundary condition $u(a) = u'(b) = 0$ or $u'(a) = u(b) = 0$ are given by

$$\mu_k = \left(\frac{\pi_p k}{b-a} \right)^p, \qquad v_k = \left(\frac{\pi_p k}{2(b-a)} \right)^p.$$

A.3.3 General Case

The eigenvalue problems (3.5)–(3.6) for general w need some heavy machinery. However, the following results also hold for-higher dimensional problems:

$$-\Delta_p u = \lambda w(x)|u|^{p-2}u, \qquad x \in \Omega \subset R^N.$$

For brevity we will focus only on one-dimensional problems, although the extension of the following results to the N-dimensional case is straightforward.

A.3.3.1 The Functional Setting

Let us define the functionals $\mathscr{A}, \mathscr{B} : W_0^{1,p}(a,b) \to R$:

$$\mathscr{A}(u) = \frac{1}{p} \int_a^b |u'|^p, \qquad \mathscr{B}(u) = \frac{1}{p} \int_a^b w(x)|u|^p.$$

Let $A : W_0^{1,p}(a,b) \to W^{-1,p'}(a,b)$ be the Gâteaux derivative of $\mathscr{A}$,

$$A(u) = -(|u'|^{p-2}u')',$$

which is an odd potential operator, uniformly continuous on bounded sets, satisfying the following condition:

$$(S) \qquad \begin{array}{l} \text{If } u_j \rightharpoonup u \text{ weakly in } W_0^{1,p}(a,b) \text{ and } A(u_j) \to u^*, \\ \text{then } u_j \to u \text{ strongly in } W_0^{1,p}(a,b) \end{array}$$

Let $B : W_0^{1,p}(a,b) \to W^{-1,p'}(a,b)$ be the odd potential operator

$$B(u) = w(x)|u|^{p-2}u,$$

which is strongly sequentially continuous, and let us observe that $\mathscr{B}(u) \neq 0$ implies that $B(u) \neq 0$.

A.3.3.2 Lyusternik–Schnirelmann Theory

Let us recall the definition of the Krasnoselskii genus $\text{gen}(C)$, where C is a closed subset of $W_0^{1,p}(a,b) \setminus \{0\}$, and C is symmetric, that is, $C = -C$.

We say that $\text{gen}(C) = n$ if n is the minimum integer such that there exists an odd continuous mapping $\varphi : C \to R^n \setminus \{0\}$.

Lyusternik and Schnirelmann proved that an even function $I : S^{N-1} \to R$ has at least N pairs of critical points. For homogeneous operators, their arguments can be applied to eigenvalue problems if we consider them as Lagrange multipliers. There are several generalizations of the Lyusternik.Schnirelmann result to the infinite-dimensional setting; see [26, 27, 68, 99].

Let us fix $m > 0$, and define the level set of $\mathscr{A}$,

$$M_m = \{u \in W_0^{1,p}(a,b) : \mathscr{A}(u) = m\}.$$

Since M_m is a bounded set, the coercivity of A implies that there exists a positive constant ρ_m such that

$$\langle A(u), u \rangle \geq \rho_m$$

for every $u \in M_m$, and $\langle \cdot, \cdot \rangle$ is the duality product between $W_0^{1,p}(a,b)$ and $W^{-1,p'}(a,b)$.

Following Amann [1], under the previous hypotheses on the functionals and their associated operators, there exists a sequence of eigenvalues:

Theorem A.5. *Let $\mathscr{A}$, A, $\mathscr{B}$, B be defined as before, and let us define*

$$\beta_k = \sup_{C \in \mathscr{C}_k, C \subset M_m} \inf_{u \in C} \mathscr{B}(u),$$

where $\mathscr{C}_k$ is the class of compact symmetric subsets of the space $W_0^{1,p}(a,b)$ of genus greater than or equal to k.

If $\beta_k > 0$, there exists an eigenfunction $u_k \in M_m$ with

$$\mathscr{B}(u) = \beta_k = m/\lambda_k.$$

Moreover, if

$$\gamma(\{w \in M_m : \mathscr{B}(u) \neq 0\}) = \infty,$$

then there exist infinitely many eigenfunctions.

We will say that $\{\lambda_k\}_k$ is the set of *variational* eigenvalues. For problems (3.5)–(3.6), it is easy to show that they exhaust the spectrum; see [48]. For the Neumann problem the same result holds, and it was proved in [95]. However, for periodic problems there exist nonvariational eigenvalues; see [5]. For higher dimensions, it is an open problem to fully characterize the spectrum. This is the main difference between one-dimensional and N-dimensional problems, and perhaps the most important open problem in this area.

It is possible to work with an equivalent characterization of the eigenvalues by introducing a Rayleigh-type quotient as in the linear case,

$$\lambda_k = \inf_{C \in \mathscr{C}_k} \sup_{u \in C} \frac{\int_a^b |u'|^p}{\int_a^b w(x)|u|^p}. \tag{3.9}$$

The equivalence follows from the homogeneity of $\mathscr{B}(u)$. Indeed, for a given $C \subset W_0^{1,p}(a,b) \setminus \{0\}$, we have a set $\tilde{C}$ in M_m by taking the retraction

$$u \mapsto \frac{u}{\mathscr{A}^{1/p}(u)}.$$

Both sets have the same genus, and

$$\inf_{u \in C} \frac{\mathscr{B}(u)}{\mathscr{A}(u)} = \inf_{u \in \tilde{C}} \mathscr{B}(u).$$

See also [53, 103] for more details.

Finally, let us observe that w can be allowed to change sign. Assuming that

$$|\Omega^+(w)| = |\{x \in (a,b) : w(x) > 0\}|_1 > 0,$$

$$|\Omega^-(w)| = |\{x \in (a,b) : w(x) < 0\}|_1 > 0,$$

where $|A|$ denotes the measure of A, there exist a sequence of positive variational eigenvalues and another sequence of negative eigenvalues.

A.3.4 Resonant Systems

We will consider the system

$$-(|u'|^{p-2}u')' = \lambda w(x)\alpha|u|^{\alpha-2}u|v|^\beta \qquad x \in (a,b),$$
$$-(|v'|^{q-2}v')' = \lambda w(x)\beta|u|^\alpha|v|^{\beta-2}v \qquad x \in (a,b), \tag{3.10}$$

with zero Dirichlet boundary conditions

$$u(a) = u(b) = v(a) = v(b) = 0,$$

where the weight $w \in L^1(a,b)$ is allowed to change sign. We assume that

$$\frac{\alpha}{p} + \frac{\beta}{q} = 1 \quad \text{and} \quad 1 < q \le p < \infty, \tag{3.11}$$

a condition that gives some homogeneity, since the solutions are invariant under the rescaling $(u,v) \to (t^{\frac{1}{p}}u, t^{\frac{1}{q}}v)$. See Boccardo and De Figueiredo [8], and the book [117] for a recent survey of different aspects of the theory of resonant systems.

The system (3.10) corresponds to the Euler–Lagrange equations of the functional

$$\Phi(u,v) = \frac{1}{p}\int_a^b |u'|^p dx + \frac{1}{q}\int_a^b |v'|^q dx - \lambda \int_a^b w(x)|u|^\alpha |v|^\beta dx, \qquad (3.12)$$

and we will say that $(u,v) \in W_0^{1,p}(a,b) \times W_0^{1,q}(a,b)$ is a solution (in the weak sense) if it is a critical point of the functional Eq. (3.12) satisfying

$$\int_\Omega |\nabla u|^{p-2}\nabla u \cdot \nabla \varphi dx + \int_\Omega |\nabla v|^{q-2}\nabla v \cdot \nabla \psi dx$$

$$= \lambda \int_\Omega r(x)\alpha |u|^{\alpha-2} u\varphi |v|^\beta dx + \lambda \int_\Omega r(x)\beta |u|^\alpha |v|^{\beta-2} v\psi dx$$

for every test pair $(\varphi, \psi) \in W_0^{1,p}(a,b) \times W_0^{1,q}(a,b)$.

We will say that $\lambda \in \mathbb{R}$ is an eigenvalue if there exists a nontrivial solution (u,v), and the eigenvalues are obtained as before:

$$\lambda_k = \inf_{C \in \mathscr{C}_k} \sup_{(u,v) \in C} \frac{\frac{1}{p}\int_a^b |u'|^p + \frac{1}{q}\int_a^b |v'|^q}{\int_a^b r(x)|u|^\alpha |v|^\beta},$$

with $C \subset W_0^{1,p}(a,b) \times W_0^{1,q}(a,b)$.

A.4 Some Properties of the Eigenvalues and Eigenfunctions

For the weighted one-dimensional p-Laplacian with Dirichlet boundary conditions we have the following result:

Theorem A.6. *Let $\{\lambda_k\}_k$ be the eigenvalues of problems (3.5)–(3.6). Then:*

1. *Every eigenfunction corresponding to the kth eigenvalue λ_k has exactly $k+1$ zeros in $[a,b]$.*
2. *For every k, λ_k is simple.*
3. *If $\lambda_k < \lambda < \lambda_{k+1}$, the only solution of problems (3.5)–(3.6) is $u \equiv 0$.*

Now we can state the classical Sturmian theorems for the p-Laplacian equation. For a proof, see [39].

Theorem A.7. *Let $w(x)$ be a continuous and positive function, and let u be a solution of*

$$-(|u'|^{p-2}u')' = w(x)|u|^{p-2}u.$$

Then the zeros of u and u' alternate.

Moreover, if v is another solution, the zeros of u and v alternate.

Theorem A.8. *Let us consider the following problems for the p-Laplacian:*

$$-(|u'|^{p-2}u')' = w(x)|u|^{p-2}u, \tag{4.13}$$

$$-(|v'|^{p-2}v')' = W(x)|v|^{p-2}v, \tag{4.14}$$

where $w(x) \le W(x)$ are positive functions. Then every solution v of Eq. (4.14) has at least one zero between two zeros of a solution u of Eq. (4.13).

Moreover, the eigenvalues of the problems

$$-(|u'|^{p-2}u')' = \lambda w(x)|u|^{p-2}u, \qquad u(a) = u(b) = 0, \tag{4.15}$$

$$-(|v'|^{p-2}v')' = \Lambda W(x)|v|^{p-2}v, \qquad v(a) = v(b) = 0, \tag{4.16}$$

satisfy

$$\Lambda_k(W) \le \lambda_k(w).$$

Let us note that the last part of Theorem 4.15 follows immediately from the variational characterization of eigenvalues given by Eq. (3.9). As a corollary, we have the following result.

Theorem A.9. *Let us assume that $0 \le w \le M$ in $(0,L)$, and let λ_1 be the first eigenvalue of*

$$-(|u'|^{p-2}u')' = \lambda w(x)|u|^{p-2}u, \qquad u(0) = u(L) = 0.$$

Then

$$\frac{\pi_p^p}{ML^p} \le \lambda_1.$$

The result follows from Theorem 4.15, with $W(x) \equiv M$, and the explicit formula for the constant coefficient case in Theorem A.4

Finally, as k goes to infinity, the asymptotic behavior of the eigenvalues is given by

$$\lambda_k \sim \frac{\pi_p^p k^p}{\left(\int_a^b w^{1/p}(x)dx\right)^p};$$

see [48] for a proof. This expression agrees with the corresponding one for the linear case; it is enough to replace $p = 2$ to get the classical one. We will use this fact at the end of Chap. 4, and it provides a motivation for Theorem C.

References

1. Amann, H.: Lusternik–Schnirelmann theory and nonlinear eigenvalue problems. Math. Ann. **199**, 55–72 (1972)
2. Anane, A.: Simplicité et isolation de la première valeur propre du p-laplacien avec poids. C. R. Acad. Sci. Paris Sér. I Math. **305**, 725–728 (1987)
3. Antman, S. S.: The influence of elasticity on analysis: modern developments. Bull. Amer. Math. Soc. **9**, 267–291 (1983)
4. Bargmann, V.: On the number of bound states in a central field of force. Proc. Nat. Acad. Sci. USA **38**, 961–966 (1952)
5. Binding, P. A., Rynne, B. P.: Variational and non-variational eigenvalues of the p-Laplacian. Journal of Differential Equations **244**, 24–39 (2008)
6. Birman, M., Solomyak, M.: On the negative discrete spectrum of a periodic elliptic operator in a waveguide-type domain, perturbed by a decaying potential. Journal d'Analyse Mathématique **83**, 337–391 (2001)
7. Birman, M., Laptev, A., Solomyak, M.: On the eigenvalue behaviour for a class of differential operators on the semiaxis, Math. Nachr. **195**, 17–46 (1998)
8. Boccardo, L., de Figueiredo, D.G.: Some remarks on a system of quasilinear elliptic equations. NoDEA Nonlinear Differential Equations Appl. **9**, 309–323 (2002)
9. Borg, G.: Uber die Stabilität gewisser Klassen von linearen Differentialgleichungen, Ark. for Matematik, Astronomi och Fysik **31**, 1–31 (1945)
10. Borg, G.: On a Liapunoff criterion of stability. Amer. J. Math. **71**, 67–70 (1949)
11. Browder, F. E.: Variational methods for nonlinear elliptic eigenvalue problems. Bull. Amer. Math. Soc. **71**, 176–183 (1965)
12. Browder, F. E.: Infinite dimensional manifolds and non-linear elliptic eigenvalue problems. Annals of Mathematics **82**, 459–477 (1965)
13. Browder, F. E.: Nonlinear eigenvalue problems and Galerkin approximations. Bull. Amer. Math. Soc. **74**, 651–656 (1968)
14. Brown, R. C., Hinton, D. B.: Lyapunov inequalities and their applications. In: Rassias, T. M. (ed.) Survey on Classical Inequalities, pp. 1–25. Springer (2000)
15. Cakmak, D., Tiryaki, A.: On Lyapunov-type inequality for quasilinear systems. Applied Mathematics and Computation **216**, 3584–3591 (2010)
16. Cakmak, D., Tiryaki, A.: Lyapunov-type inequality for a class of Dirichlet quasilinear systems involving the (p1, p2,..., pn)-Laplacian. Journal of Mathematical Analysis and Applications **369**, 76–81 (2010)
17. Calogero, F.: Necessary Conditions for the Existence of Bound States. Nuovo Cimento **36**, 199–201 (1965)
18. Calogero, F.: Upper and lower limits for the number of bound states in a given central potential. Communications in Mathematical Physics **1**, 80–88 (1965)

J.P. Pinasco, *Lyapunov-type Inequalities: With Applications to Eigenvalue Problems*,
SpringerBriefs in Mathematics, DOI 10.1007/978-1-4614-8523-0,
© Juan Pablo Pinasco 2013

19. Cañada, A., Villegas, S.: Lyapunov inequalities for Neumann boundary conditions at higher eigenvalues. J. European Math. Soc. **12**, 163–178 (2010)
20. Cañada, A., Villegas, S.: Stability, resonance and Lyapunov inequalities for periodic conservative systems. Nonlinear Analysis: Theory, Methods, and Applications **74**, 1913–1925 (2011)
21. Cañada, A., Villegas, S.: An applied mathematical excursion through Lyapunov inequalities, classical analysis and differential equations. SeMA Journal **57** (2012)
22. Cañada, A., Montero, J.A., Villegas, S.: Lyapunov inequalities for partial differential equations. J. Functional Analysis **237**, 17–193 (2006)
23. Castro, M. J., Pinasco, J.P.: An Inequality for Eigenvalues of Quasilinear Problems with Monotonic Weights. Applied Mathematical Letters **23**, 1355–1360 (2010)
24. Cheng, S. S.: A discrete analogue of the inequality of Lyapunov. Hokkaido Math. J. **12**, 105–112 (1983)
25. Cheng, S. S.: Lyapunov inequalities for differential and difference equations. Fasc. Math. **23**, 25–41 (1991)
26. Clark, D. C.: A variant of the Ljusternik–Schnirelmann theory. Indiana Univ. Math. J. **22** 65–74 (1972)
27. Coffman, C. V.: A minimum–maximum principle for a class of nonlinear integral equations. J. d'Analyse Mathématique **22**, 391–419 (1969)
28. Cohn, J. H. E.: On the Number of Negative Eigen-Values of a Singular Boundary value Problem. J. London Math. Soc. **40**, 523–525 (1965)
29. Cohn, J. H. E.: On the Number of Negative Eigen-Values of a Singular Boundary Value Problem (II). Journal of the London Mathematical Society, **41**, 469–473 (1966)
30. Cohn, J. H. E.: Consecutive zeroes of solutions of ordinary second order differential equations. J. London Math. Soc. **5**, (1972) 465–468.
31. Courant, R., Hilbert, D,: Methods of Mathematical Physics, vol. I, Interscience Publishers, Inc., New York (1953)
32. Cuesta, M.: Eigenvalue problems for the p-Laplacian with indefinite weights. Electron. J. Differential Equations **2001**, nr. 33 (2001)
33. Das, K. M., Vatsala, A. S.: Green's function for n-n boundary value problem and an analogue of Hartman's result. Journal of Mathematical Analysis and Applications **51**, 670–677 (1975)
34. De Figueiredo, D., Gossez, J. P.: On the first curve of the Fučík spectrum of an elliptic operator. Differential and Integral Equations **7** 1285–1302 (1994)
35. De Nápoli, P. L., Pinasco, J. P.: A Lyapunov Inequality for Monotone Quasilinear Operators. Differential Integral Equations **18**, 1193–1200 (2005)
36. De Nápoli, P. L., Pinasco, J. P.: Estimates for Eigenvalues of Quasilinear Elliptic Systems, Part I. Journal of Differential Equations **227**, 102–115 (2006)
37. De Nápoli, P. L., Pinasco, J. P.: Lyapunov-type inequalities in R^N. Preprint (2012)
38. del Pino, M., Drabek, P., Manásevich, R.: The Fredholm alternative at the first eigenvalue for the one-dimensional p-Laplacian. J. Differential Equations **151**, 386–419 (1999)
39. Dosly, O., Rehak, P.: Half-Linear Differential Equations. Volume 202 North-Holland Mathematics Studies. North Holland (2005)
40. Drabek, P., Kufner, A.: Discreteness and simplicity of the spectrum of a quasilinear Sturm–Liouville-type problem on an infinite interval. Proceedings of the American Mathematical Society **134**, 235–242 (2006)
41. Drabek, P., Manásevich, R.: On the closed solutions to some nonhomogeneous eigenvalue problems with p-Laplacian. Differential Integral Equations **12**, 773–788 (1999)
42. Egorov, Y. V., Kondriatev, V. A.: On Spectral Theory of Elliptic Operators (Operator Theory: Advances and Applications) Birkhäuser (1996)
43. Elbert, A.: A half-linear second order differential equation. Colloq. Math. Soc. Janos Bolyai **30**, 158–180 (1979)
44. Elias, U.: Singular eigenvalue problems for the equation $y^n + \lambda p(x)y = 0$, Monatsh. Math. **142**, 205–225 (2004)
45. Eliason, S. B.: Liapunov type inequalities for certain second order functional differential equations. SIAM J. Applied Math. **27**, 180–199 (1974)

46. Eliason, S. B.: Distance between zeros of certain differential equations having delayed arguments. Annali di Matematica Pura ed Applicata **106**, 273–291 (1975)
47. Evans, L. C.: Partial Differential Equations, American Mathematical Society, New York (2010)
48. Fernández Bonder, J., Pinasco, J. P.: Asymptotic Behavior of the Eigenvalues of the One Dimensional Weighted p-Laplace Operator. Ark. Mat. **41**, 267–280 (2003)
49. Fernández Bonder, J., Pinasco, J. P., Salort, A. M.: A Lyapunov-type Inequality and Eigenvalue Homogenization with Indefinite Weights. Preprint (2013)
50. Fernández Bonder, J., Pinasco, J. P., Salort, A. M.: Quasilinear eigenvalues problems. In print (2013)
51. Fink, A. M., St. Mary, D. F.: On an inequality of Nehari. Proceedings of the American Mathematical Society **21**, 640–642 (1969)
52. Fučík, S., Nečas, J., Souček, J., Souček, V.: Spectral analysis of nonlinear operators. Lecture Notes in Mathematics **346**. Springer-Verlag, Berlin, 1973.
53. García Azorero, J., Peral Alonso, I.: Existence and Nonuniqueness for the p-Laplacian: Nonlinear Eigenvalues. Communications in Partial Differential Equations **12**, 1389–1430 (1987)
54. García-Huidobro, M., Manásevich, R., Zanolin, F.: A Fredholm-like result for strongly nonlinear second order ODEs. J. Differential Equations **114**, 132–167 (1994)
55. García-Huidobro, M., Le, V. K., Manásevich, R., Schmitt, K.: On principal eigenvalues for quasilinear elliptic differential operators: an Orlicz–Sobolev space setting. Nonlinear Differential Equations Appl. **6**, 207–225 (1999)
56. Gossez, J. P., Manásevich, R.: On a nonlinear eigenvalue problem in Orlicz–Sobolev spaces. Proc. Roy. Soc. Edinburgh A **132**, 891–909 (2002)
57. Guedda, M., Veron, L.: Bifurcation Phenomena Associated to the p-Laplace Operator. Transactions of the American Mathematical Society **310**, 419–431 (1988)
58. Guseinov, G.Sh., Kaymakcalan, B.: On a disconjugacy criterion for second order dynamic equations on time scales. J. of Comp. and Appl. Math. **141**, 187–196 (2002)
59. Ha, C.-W.: Eigenvalues Of A Sturm–Liouville Problem and Inequalities of Lyapunov Type. Proceedings of the American Mathematical Society **126**, 3507–351 (1998)
60. Hardy, G. H., Littlewood, J. E., Pólya, G.,: Inequalities. Cambridge University Press, London (1988)
61. Harris, B. J.: On an inequality of Lyapunov for disfocality. Journal of Mathematical Analysis and Applications **146**, 495–500 (1990)
62. Harris, B. J., Kong, Q.: On the oscillation of differential equations with an oscillatory coefficient. Transactions of the American Mathematical Society **347**, 1831–1839 (1995)
63. Hartman, P., Wintner, A.: On an Oscillation Criterion of Liapunoff. American Journal of Mathematics **73**, 885–890 (1951)
64. Hille, E.: An Application of Prüfer's Method to a Singular Boundary Value Problem. Mathematische Zeitschrift **72**, 95–106 (1959)
65. Hochstadt, H.: On an inequality of Lyapunov. Proceedings of the American Mathematical Society **22**, 282–284 (1969)
66. Hong, H-L., Lian, W-C., Yeh, C.C.: The oscillation of half-linear differential equations with an oscillatory coefficient. Mathematical and Computer Modelling **24**, 77–86 (1996)
67. Hudzik, H., Maligranda, L.: Some remarks on submultiplicative Orlicz functions. Indagationes Mathematicae **3**, 313–321 (1992)
68. Jabri, Y.: The Mountain Pass Theorem, Variants, Generalizations and Some Applications. Encyclopedia of Mathematics and Its Applications, Cambridge University Press (2003)
69. Kabeya, Y., Yanagida, E.: Eigenvalue problems in the whole space with radially symmetric weight. Communications in Partial Differential Equations **24**, 1127–1166 (1999)
70. Kac, M.: Can One Hear the Shape of a Drum?, American Math. Monthly (Slaught Mem. Papers, nr. 11) **73**, 1–23 (1966)
71. Kolodner, I.: Heavy rotation string: a nonlinear eigenvalue problem. Comm. Pure Appl. Math. **8**, 395–408 (1955)

72. Kusano T., Naito, M.: On the Number of Zeros of Nonoscillatory Solutions to Half-Linear Ordinary Differential Equations Involving a Parameter. Transactions of the American Mathematical Society **354**, 4751–4767 (2002)

73. Kwong, M. K.: On Lyapunov's inequality for disfocality. J. Math. Anal. Appl. **83**, 486–494 (1981)

74. Lazer, A. C., McKenna, P. J.: Large-amplitude periodic oscillations in suspension bridges: some new connections with nonlinear analysis. SIAM Review **32**, 537–578 (1990)

75. Lee, C.-F., Yeh, C.-C., Hong, C.-H., Agarwal, R. P.: Lyapunov and Wirtinger Inequalities. Appl. Math. Letters **17**, 847–853 (2004)

76. Levin, A.: A Comparison Principle for Second Order Differential Equations. Sov. Mat. Dokl. **1**, 1313–1316 (1960)

77. Lewis, J. L.: Uniformly fat sets. Transactions of the American Mathematical Society **308**, 177–196 (1988)

78. Li, H. J., Yeh, C. C.: Sturmian Comparison Theorem for Half Linear Second Order Differential Equations. Proc. Royal Soc. Edinburgh, Sect. A **125**, 1193–1204 (1995)

79. Lützen, J., Mingarelli, A.: Charles François Sturm and Differential Equations. In: Pont, J.-C. (ed.) in collaboration with Padovani F.: Collected Works of Charles François Sturm, pp. 25–48. Birkäuser (2009)

80. Lyapunov, A.: *Problème General de la Stabilité du Mouvement*, Ann. Math. Studies 17, Princeton Univ. Press, 1949 (reprinted from Ann. Fac. Sci. Toulouse, **9**, 204–474 (1907). Translation of the original paper published in Comm. Soc. Math. Kharkow (1892)

81. Makai, E.: Über die Nullstellen von Funktionen, die Lösungen Sturm–Liouville'scher Differentialgleichungen sind. Comment. Math. Helv. **16**, 153–199 (1944)

82. Merdivenci Atici, F., Guseinov, G. Sh., Kaymakcalan, B.: On Lyapunov Inequality in Stability Theory for Hills Equation on Time Scales. J. of Ineq. and Appl. **5**, 603–620 (2000)

83. Mustonen, V., Tienari, M.: An eigenvalue problem for generalized Laplacian in Orlicz–Sobolev spaces. Proc. Royal Soc. Edinburgh A **129**, 153–163 (1999)

84. Naimark, K., Solomyak, M.: Regular and pathological eigenvalue behavior for the equation $-\lambda u'' = Vu$ on the semiaxis. J. Functional Analysis **151**, 504–530 (1997)

85. Naito, M.: On the Number of Zeros of Nonoscillatory Solutions to Higher-Order Linear Ordinary Differential Equations. Monatsh. Math. **136**, 237–242 (2002)

86. Nehari, Z.: On the zeros of solutions of second-order linear differential equations. American Journal of Mathematics **76** 689–697 (1954)

87. Nehari, Z.: Oscillation criteria for second-order linear differential equations. Transactions of the American Mathematical Society **85**, 428–445 (1957)

88. Nehari, Z.: Some eigenvalue estimates. Journal d' Analyse Mathematique **7**, 79–88 (1959)

89. Nehari, Z.: Extremal problems for a class of functionals defined on convex sets. Bull. Amer. Math. Soc. **73**, 584–591 (1967)

90. Osserman, R.: A note on Hayman's theorem on the bass note of a drum. Comment. Math. Helv. **52**, 545–555 (1977)

91. Pachpatte, B. G.: Lyapunov-type integral inequalities for certain differential equations. Georgian Math. J. **4**, 139–148 (1997)

92. Pachpatte, B. G.: Mathematical Inequalities. North Holland Math. Library, Elsevier (2005)

93. Patula, W. T.: On the Distance between Zeroes. Proceedings of the American Mathematical Society **52**, 247–251 (1975)

94. Pinasco, J. P.: Lower bounds for eigenvalues of the one-dimensional p-Laplacian. Abstract and Applied Analysis **2004**, 147–153 (2004)

95. Pinasco, J. P.: Lower bounds of Fučík eigenvalues of the weighted one dimensional p-Laplacian. Rendiconti dell'Inst. Matematico dell'Univ. di Trieste XXXVI, 49–64 (2004)

96. Pinasco, J. P.: The Distribution of Non-Principal Eigenvalues of Singular Second Order Linear Ordinary Differential Equations. Int. J. of Mathematics and Mathematical Sciences **2006**, 1–7 (2006)

97. Pinasco, J. P.: Comparison of eigenvalues for the p-Laplacian with integral inequalities, Appl. Math. Comput. **182**, 1399–1404 (2006)

98. Pinasco, J. P., Scarola, C.: Density of Zeros of Eigenfunctions of Singular Sturm Liouville Problems. Preprint (2013)
99. Rabinowitz, P. H.: Minimax methods in critical point theory with applications to differential equations. Conference Board of the Mathematical Sciences, Amer. Math. Soc. (1986)
100. Reid, W. T.: A matrix Liapunov inequality. J. of Math. Anal. and Appl. **32**, 424–434 (1970)
101. Reid, W. T.: A Generalized Lyapunov Inequality. J. Differential Equations **13**, 182–196 (1973)
102. Reid, W. T.: Interrelations between a trace formula and Liapunov type inequalities. J. of Differential Equations **23**, 448–458 (1977)
103. Riddell, R. C.: Nonlinear eigenvalue problems and spherical fibrations of Banach spaces. J. of Functional Analysis **18**, 213–270 (1975)
104. Sanchez, J., Vergara, V. A Lyapunov-type inequality for a ψ-Laplacian operator. Nonlinear Analysis **74**, 7071–7077 (2011)
105. Solomyak, M.: On a class of spectral problems on the half-line and their applications to multi-dimensional problems. Preprint arXiv:1203.1156 (2012)
106. St. Mary, D. F.: Some Oscillation and Comparison Theorems for $(r(t)y')' + p(t)y = 0$, J. of Differential Equations, **5**, 314–323 (1969)
107. Sturm, C.: Mémoire Sur les Équations différentielles linéaires du second ordre; Journal de Liouville **I**, 106–186 (1836). In: Pont, J.-C. (ed.) in collaboration with Padovani F.: Collected Works of Charles François Sturm. Birkäuser (2009).
108. Tang, X. H., He, X.: Lower bounds for generalized eigenvalues of the quasilinear systems. Journal of Mathematical Analysis and Applications **385**, 72–85 (2012)
109. Tienari, M.: Lyusternik–Schnirelmann Theorem for the Generalized Laplacian. J. Differential Equations **161**, 174–190 (2000)
110. Watanabe, K.: Lyapunov-type inequality for the equation including 1-dim p-Laplacian. Mathematical Inequalities and Applications **15**, 657–662 (2012)
111. Watanabe, K., Kametaka, Y., Yamagishi, H., Nagai, A., Takemura, K.: The best constant of Sobolev inequality corresponding to clamped boundary value problem. Boundary Value Problems **2011** Article ID 875057 (2011)
112. Weyl, H.: Über die asymptotische Verteilung der Eigenwerte, Nachrichten der Königlichen Gesellschaft der Wissenschaften zu Göttingen, 110–117 (1911)
113. Wintner, A.: On the Non-Existence of Conjugate Points. American Journal of Mathematics **73**, 368–380 (1951)
114. Yang, X.: On inequalities of Lyapunov type. Applied Math. Comp. **134**, 293–300 (2003)
115. Yang, X., Kim, Y.-I., Lo, K.: Lyapunov-type inequality for a class of quasilinear systems. Mathematical and Computer Modelling **53**, 1162–1166 (2011)
116. Zhang, Q.-M., Tang, X. H.: Lyapunov-type inequalities for even order difference equations. Applied Mathematics Letters **25**, 1830–1834 (2012)
117. N.B. Zographopoulos (ed.) Estimates for Eigenvalues of Quasilinear Elliptic Systems. Scienpress Ltd. (2012)